Green Energy and Technology

More information about this series at http://www.springer.com/series/8059

Guedi Capeluto · Carlos Ernesto Ochoa

Intelligent Envelopes for High-Performance Buildings

Design and Strategy

Springer

Guedi Capeluto
Faculty of Architecture and Town Planning
Technion IIT
Haifa
Israel

Carlos Ernesto Ochoa
Faculty of Architecture and Town Planning
Technion IIT
Haifa
Israel

ISSN 1865-3529 ISSN 1865-3537 (electronic)
Green Energy and Technology
ISBN 978-3-319-39254-7 ISBN 978-3-319-39255-4 (eBook)
DOI 10.1007/978-3-319-39255-4

Library of Congress Control Number: 2016956188

Printed on acid-free paper

This Springer imprint is published by Springer Nature
The registered company is Springer International Publishing AG
The registered company address is: Gewerbestrasse 11, 6330 Cham, Switzerland

Preface

The role of the building envelope is primarily to act as a barrier between inside and outside and, in this way, provide privacy and protection for its inhabitants from external environmental factors (sun, wind, and rain), while retaining thermal comfort, a healthy living environment and the ability to see outside. One may, moreover, argue that the building envelope serves as a mediator, reconciling inside and outside, so that the building's relationship to its nearby surroundings is established.

Indeed, one may observe throughout history the manner and degree to which such functions have been changed and upgraded in response to emerging technologies and materials. Alongside newly defined needs, building envelopes have to answer to changing demands for comfort and occupants' relation to their environment.

This book was motivated by the worrying realization that often, architects overlook some of the original functions of the building skin, passing very late in the design process the responsibility of its performance to mechanical and building envelope engineers. It is vital to understand the significance of design decisions made from the very beginning of the process about building form and envelope in order to achieve intelligent envelopes for high-performance buildings. Understanding their constituent layers and materials, and how the entire building relates to its outside, signals the hallmark of sustainable architectural solutions.

Under such context, this book lays out basic principles of climatic design for high-performance building envelopes, followed by practical examples and design tools that can be used to analyze and obtain a series of adequate high-performance solutions. The book condenses an ample body of knowledge based on years of research and presents useful guidance for designers that want to apply novel technologies for the design of intelligent envelopes in a suitable and practical way. It also addresses new challenges for the twenty-first century, encouraging energy conservation and emission reduction in the construction of new buildings, as well as renovation of existing ones toward sustainable reuse.

It is paramount that architects take on again the responsibility for envelope design as part of the whole architectural design process, as an integral part of the interactive space between building and its environment. As we show in our book, the introduction of new technologies and materials presents a unique opportunity. It calls for an innovative rethinking of the role played by the building envelope, so as to create healthy building environments that relate well to the outside and succeed in making better use of natural resources, such as sun, light, and air.

Haifa, Israel Guedi Capeluto

Contents

Chapter 1
What Is a Real Intelligent Envelope?

1.1 Definition of Intelligence

In a colloquial context, the term "intelligence" is used to define cleverness or as a synonym of smartness. One might even hear about "intelligent buildings" or "smart windows", when applied to inanimate objects that seem to be doing something complex or with some similarity to a human thought process.

Many dictionary meanings (Merriam-Webster 2015; Oxford 2015) give another direction to this word. They define it as an ability. This skill allows to understand a given situation and apply a response to achieve a desired performance according to the specific starting condition.

This makes intelligence an intrinsic characteristic of all living beings, even if they employ basic mechanisms such as single physical or chemical reactions in order to ensure their survival. But the processes do not occur randomly or without any balance. It is interesting to note that intelligent mechanisms in nature must respect equilibrium of certain variables, such as their overall dimensions according to species, and energy consumption requirements in order for them to continue to be viable (Britannica Online 2015).

Therefore, true intelligence is sustainable by definition. Aiming to a correct application of the concept to the design of our living spaces, opens up possibilities of producing high-performance buildings that "know" by themselves and with little intervention how to utilize only the minimum required energy. To achieve it, they would make, by themselves or through appropriate reactions, the necessary adjustments to provide comfort to its occupants, and at the same time keep balance with its immediate environment.

Intelligence in nature is also adaptable, using a variety of mechanisms or methods to provide different responses that restore equilibrium to the organism. Some of these mechanisms are apparently activated without the intervention of conscious decisions. One such example can be seen in the skin. When cold weather is sensed, there is involuntary shivering and blood vessels carry less blood, in order

G. Capeluto and C.E. Ochoa, *Intelligent Envelopes for High-Performance Buildings*, Green Energy and Technology, DOI 10.1007/978-3-319-39255-4_1

to increase body temperature by producing some heat and preserving existing energy. On the other hand, sweating is produced involuntarily and pore dilation is seen when hot weather is detected, lowering temperature through evaporation (Tortora and Derrickson 2014).

The performance of an intelligent organism requires being composed of intelligent components that contribute to the main aim of survival and development. In this book we focus on the outer layer of a building, or envelope. In most cases it is identified with the roof and facades, even though in some recent cases of contemporary architecture their distinction is blurred.

The building envelope is one of the most important components of any habitable structure, as it protects occupants from external weather, contributes a comfortable interior, allows communication if needed with the exterior, and last but not least important, provides the definite impression and image we get for our perception of "a building".

Therefore, an intelligent building envelope will be understood as the outer layer of a building, designed through a specific process for adaptability to the challenges posed by interior and exterior conditions using minimum energy. This process has also provided the elements and mechanisms that follow the stages of intelligence, carrying out the adaptation strategies for which it has been planned.

1.2 Stages of the Intelligent Process

Different disciplines have studied the subject of intelligence, providing different perspectives and manners to analyze it as a process. Scientists and philosophers alike have taken a series of approaches towards explaining how it works, based on observations from nature and experimentation. Many prevalent theories on the phases of intelligence draw from biological studies on how living beings acquire information and respond to it.

A disadvantage of a solely biological approach is that conclusions found in many studies cannot be generalized to all processes and types of mechanisms, including those performed by inanimate objects.

An alternative approach, with a high degree of complexity, constructs its stages based solely on the human intelligence process. There is also no general agreement among researchers, since many steps are conditioned by later learning from the individual, or by the society where human activity takes place. The stages need to be generalized to such level that inanimate objects can also perform them using a series of even simpler steps.

To overcome these shortcomings, the steps that compose the intelligence process will be studied from a functional point of view. An advantage of the approach is that functions are able to be emulated and replicated in different contexts, such as computer simulation. Even though, it must be pointed out that results can only be similar as long as the circumstances and methods being emulated are close enough to the original case.

There are different points of view even within the functional school of thought for intelligence. Some methodologies try to integrate a psychological point of view, with the aim of ultimately emulating human intelligence. Others follow a more biological path, also with the objective of producing artificial entities that are close to what we know in nature. In this book we shall take a middle approach, since intelligence for buildings still requires long strides towards a successful application, but we still want the field to advance beyond the current state of the art.

A classical proposition to examine an intelligent process, using a functional approach based on biological and psychological observations, divides it into three main linear stages (Hayes-Roth 1995):

$$Perception \rightarrow reasoning \rightarrow action$$

Throughout this book, the steps mentioned in this approach will be used. These divisions have been recommended for use in research involving intelligence emulation for computer science and robotics. They are also suitable for application to other fields of study such as architectural design and planning. In this field, using the divisions enables adapting the creative process of building envelopes to be conscious about following suitable steps for a given objective (in our case, high-performance through adequate climate approaches). The approach also provides a different viewpoint from what is traditionally available in design.

Let us take a detailed look at each of the steps of the intelligence process according to this approach. On their own, the stages can have a series of sub-processes to carry them out successfully. Specialized literature exits on each of the steps, therefore we summarize the main points relevant to our architectural design approach. For simplicity, we shall omit any secondary connection processes that link between the main stages, but which are needed for successful communication and reaction.

The first stage, *perception*, deals with the acquisition of information about the surrounding environment (Corsini 1987) and its transformation into the communication format that the intelligent being uses for collecting such information.

During this stage, information gained is also organized, identified and interpreted (Schacter et al. 2011) for use in the second stage. The physical environment is mapped and its properties are recognized, either by using rules or neural structures that communicate these properties (Bach 2009). A certain "consciousness" is expressed by recognition of more abstract variables such as "space" and "time", which go beyond simple information collection.

In nature, this information procurement is done through different biological receptors, which in humans are generally identified with the five senses. The receptors are tissues ("sensors"), which translate the stimulus caused by particles of light, air, liquid or air movements and pressure into distinctive signals. These signals afterwards become electrical impulses distinctly understood and recognized by the brain as sight, sound, smell, taste or touch. Figure 1.1 shows a schematic diagram showing how perception is understood.

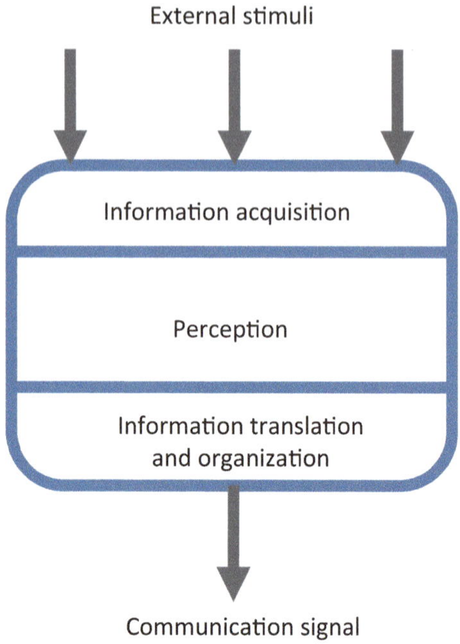

Fig. 1.1 Perception as information acquisition

The second stage, *reasoning*, is more complex to describe as it is usually confused with intelligence itself. Classical definitions state it as a problem-solving procedure in which solutions are logically deduced (Reber 1985), as well as through the application of a set of pre-established rules similar to logical postulates (O'Brien 2009).

For the purposes of our subject, this explanation will be used. Nevertheless, it is acknowledged that such definition implies a rationalized and pre-defined set of actions and priorities in order to reach a solution. In other words, a decision derived out of consciousness. Nevertheless, it has to be pointed out that "consciousness" is more of a human-related term that expresses self-identity, as well as identification with abstract concepts such as values and goals. Consciousness in nature is not well understood, as living beings seen from a human viewpoint, seem to have as main objective their survival.

These sets of rules, unwritten as they are, can be seen transmitted from one generation to the other in nature, such as animals teaching their offspring how to hunt. Or they seem to be embodied within their neural systems, for example when specific flocks of birds decide to migrate at a certain time, while others of the same species stay in a similar place for some days longer. In our case, we shall also consider the case of rules being placed by an intelligent being into a non-intelligent one, such as the programming of a computer by a human operator.

On the other hand, many times it appears that reasoning in nature occurs in what could be called an automated manner (such as reflexes), or in a way that is too fast to record its steps, reasoning or rules. Higher level intelligence systems are able to construct new sets of rules, or modify their existing ones, according to experience (Fagin et al. 2004).

For reasoning to take place, translated signals are received from the receptors that indicate a given external condition. According to the complexity of the reasoning system, data can be *interpreted* on different levels, such as a series of environmental data (geographical location, time, heat, humidity, wind direction, etc. without internal meaning) or as a situation that has some meaning to the reasoning being (12 o'clock in London on a rainy day).

Logical rules might indicate adopting an almost automatic action (take cover from rain to keep dry), while deduction from the given situation might refine the functionality and allow further experiences (take cover from the rain inside a restaurant as it is close to lunch hour) (an example adapted from Goel 2007). Presently, inanimate objects comply only with the logical rules, while being devoid of understanding the enjoyment of deductive ones!

Figure 1.2 shows a scheme of how reasoning uses rules to interpret the sensory information provided by perception.

Action, the third step, is the triggering or inhibition of an activity as response to the initial stimulation (Corsini idem) that was detected by the perception system. It executes the conclusions reached by the logical system, carrying out the desired objective in a noticeable form, performing a change in the physical properties of the intelligent being. For such event to take place, the reasoning center must be in constant communication with specific organs or tissues that carry out the desired function. It should also know which organ to send information and what kind of instruction to be sent.

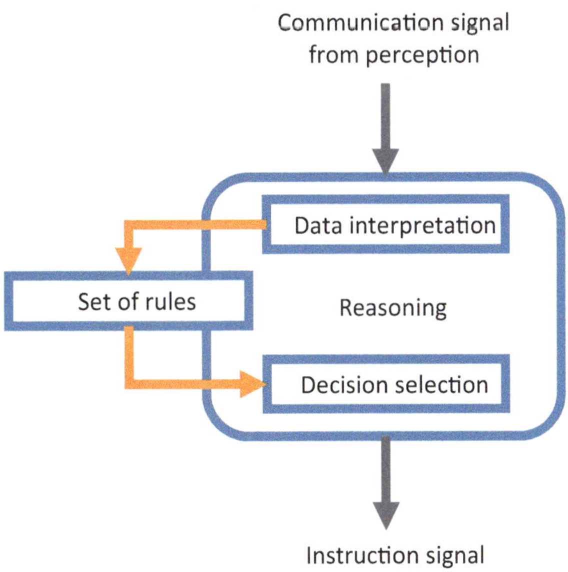

Fig. 1.2 Functional interpretation of reasoning using a set of rules

Such response would give as consequence a change in the internal state of the living being (for example, an increase in blood pressure as response to lowering external temperatures), or would be expressed as a modification on its immediate surrounding environment (such as the construction of nests and burrows to accommodate for newborn species).

Signals are sent from the reasoning center which are interpreted by cells, tissues, organs, limbs, etc., as instructions to perform a specific action (Rowland 1992). Different mechanisms are available in nature, either physical, chemical, or a complete series of activities. For humans, actions would involve starting a series of noticeable events both physical and intellectual in order to reach a goal. In nature it is observed that the set of actions has some delimitation according to the species, meaning that we do not observe beings taking actions that correspond to other beings.

A schematic representation of action defined as triggering of an activity is provided in Fig. 1.3.

It must be noted that, despite the enumeration of three steps with a clear start and end, intelligent processes in general are not linear when observed in real situations. Organisms know when to stop, start and combine their processes as required. They would not lock themselves in a single one, or stop performing a certain action as suggested by the single-end procedure approach.

Natural intelligent systems have *feedback mechanisms* that not only repeat or modify the cycle if required, but maintain the overall internal equilibrium of the

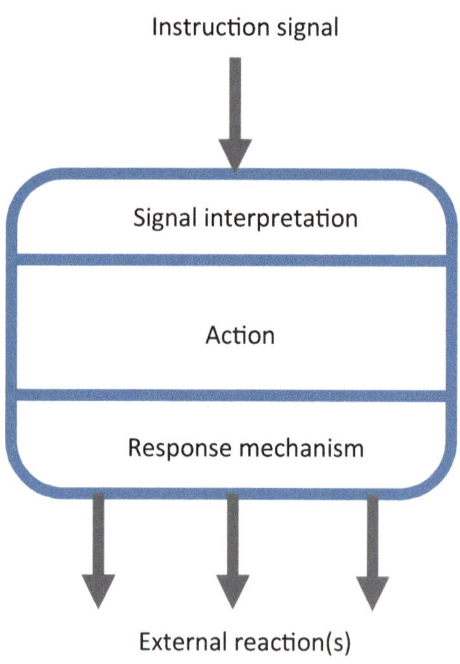

Fig. 1.3 Action as the triggering of an activity

living being, so that under normal circumstances no process would overcome or take more energy than required for other processes.

Feedback mechanisms might activate other procedures to restore equilibrium. For example, the thermostat that regulates internal temperature of a higher level organism, can decide if cooling or heating procedures must be engaged by the tissues. This search for the stability state is called homeostasis (Tortora and Derrickson, idem). In other words, homeostasis makes sure that the whole and its parts, such as the outer envelope of an organism, contribute to the common goal of organism survival and development.

As suggested by the steps and feedback mechanisms, intelligence is a concept with inherent dynamism and long-term adaptability towards maintaining a given goal. Changing conditions are answered by a set of responses that are adequate for each situation. These responses do not contradict each other, use the least energy possible and contribute to the general well-being of the organism. They are always being adjusted according to information provided by overall external conditions. This makes the emulation of intelligence a challenge if we are to achieve sustainable design in building envelopes.

A summary of the intelligence steps according to the functional approach, and their main contributions to the process is given in Fig. 1.4. It shows the process as a linear one with the included addition of the feedback mechanism.

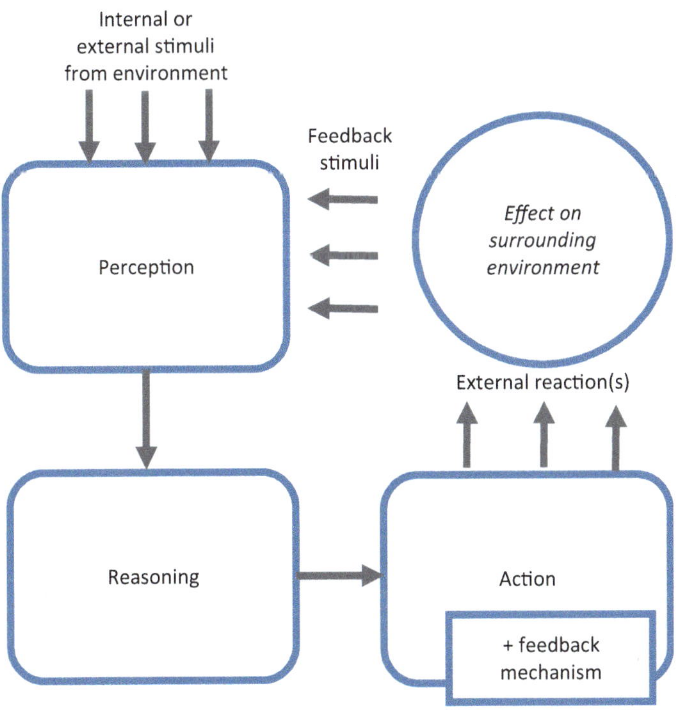

Fig. 1.4 Summary of intelligence steps according to a classical functional approach, with the addition of feedback

1.3 Applying Artificial Intelligence to Objects

The previous section has shown how intelligence is present in living beings and helps them keep their natural equilibrium with a series of adequate responses that are chosen from an unwritten set of rules or logical deductions. These responses and the underlying mechanisms respect energy limits, according to the size of each being and allow endurance to a certain range of conditions.

However, some of these beings also construct structures (objects) that are not part of their bodies. They are used to live and develop inside them, such as nests or burrows. They serve to modify the immediate environment for the benefit of the living creature, and also aid to survive beyond the original range of conditions that are endurable by that living being without the constructed object.

Another fact that should be pointed out, is that these objects have been built intelligently with some degree of complexity, using local resources, and commonly do not use any energy at all.

Humans, on the other hand, also produce structures and buildings in order to shelter themselves from adverse climate and to modify their environment. Yet, traditional design methods and materials used in construction allow only a fixed type of building usage. Many times, occupancy in buildings can be done at the expense of high energy expenditure and low adaptability.

Most human-made constructions usually contradict the postulates of intelligence, dynamism and adaptability presented in the previous section. This presents a paradox of an intelligent being not producing intelligent objects and even entering in conflict with the resources offered by its surroundings.

Can buildings or their parts be made intelligent?

The answer is yes. Intelligence as a concept can be extended to the design of products created by humans. Human-made objects potentially can also be intelligent themselves, in the degree that their pertinent technological implementation allows replication of the intelligence process.

It must be remembered that intelligence in nature can replicate itself. For example, offspring are born inheriting certain traits that are also characteristic to the parents. By some inference process, we can deduct that under normal circumstances, the method to generate and choose intelligent building elements (such as envelopes) has to be intelligent too in order to preserve the same degree of abilities as the being that created it.

The method has to understand and be conscious about the intelligent processes that allow the achievement of an objective, choosing the best elements for that aim. More will be elaborated on the subject in the following chapters.

If we are to use the concept of intelligence in architecture, then each element of the building should be extensions of their sustainable adaptability. This is particularly true if architecture is produced in order to provide among other things, protection and comfort. The concept of intelligent elements in buildings can bring substantial benefits in terms of energy savings, adaptability to changing situations and overall user satisfaction.

Intelligence process for machines

Machines are systems composed of different parts that use energy with the purpose of aiding in performing a certain task (AHD 2016). It has been the aim of different researchers to provide these inanimate objects with intelligence, in order to make them more efficient or even to imitate human reasoning functions and activities. Emulating intelligence through machines has a long history rooting into antiquity (Poole et al. 1998). Due to their old age, many details are not available, but many early examples consist of objects using ingenious mechanisms. They could do a series of repetitive, sometimes interactive, simple tasks which provided spectators with the feeling of watching a sentient being.

The intelligence process for machines can also be analyzed using a functional approach. As such, there can be a correspondence between the steps of the intelligent process theorized for living beings (perception, reasoning and action) and the functions that need to be fulfilled by machines. In order to reach this goal, mechanical actions and technological means have to be adequate, performing according to these divisions.

Perception in living beings is generally carried out through sensors made out of cells, residing on the boundary between inside and outside, translating different stimuli into signals. They can also be composed of specialized bio-materials that are integrated to the processing centers of simpler animals or plants, reacting and providing the simple being with information.

Translating input from the real world to machines is a non-trivial task, which forms a field of study in itself. Different degrees of complexity can be found on this translation chore, and reflect somewhat the evolution of how autonomous machines can become. All of these methods have as objective translating real world information into knowledge, as data that can be simplified for analysis as smaller parts of the bigger whole.

The most rudimentary way is to use mechanical means such as levers or wheels to produce similarly mechanical reactions, such as in the first calculators ever produced. Nowadays, we might find an easier method but with the same intentions, when human operators enter data as required into a software program through a computer keyboard. In both cases the formats might differ in great amount from that used by daily human logic. Translation is mostly made at the point of the data input, as electrical signals representing information.

Another example is the transformation of natural stimuli into electricity using different materials and their physical properties. Examples include thermometers, light receptors, pressure recorders, etc. Such method facilitates data collection for a diversity of variables and is widely used today. It emulates, in an artificial way, biological receptors such as the eye, and provide machines with an array of information for processing. In this way, we can talk of artificial *sensors* carrying out the perception for machines.

On a more sophisticated extreme, futuristic proposals include new materials combining organic and mechanical receptors. It also includes custom-made materials where their molecular properties have been modified, according to different objectives and converting energy from one form to the other, beyond electricity (Janocha 2007).

The reasoning stage, which distinguishes living organisms, is carried out in machines by rather elementary means. First, exclusively mechanical processes were used in which a given physical condition (temperature, weight) triggered another mechanism that would provide a reaction interpreted as an answer by the human operator. Today it is more common to find electronic programmatic processes in almost every machine. Programmatic methods use mostly a "yes-no" gate logic for directing electrical impulses and performing calculations. Signals received from sensors are interpreted into corresponding information for use in a logic system through knowledge representation.

The processing of this information is what makes a machine "intelligent" and different from an equivalent simpler machine, or a set of machines, performing similar functions. Information processing is usually done through a set of instructions, which also lead to subsequent rules in terms of what has to be done based on the information received. Imbuing logic to an inanimate object is one of the central themes of research for intelligent behavior in machines. Approaches differ if this logic should be based on knowledge or signals, or to compose something similar to thinking (Nilsson 1998).

In complex intelligent systems, processing is concentrated in a *processor*, or control unit. Informally people refer to this center as the "brain" of the machine, since it emulates the functions performed by the central processing organ in many living beings.

The third step, action, is carried out in machines by specialized elements called *actuators*. They are meant to produce the physical or electronic reaction decided by the processing system to a given condition. It can vary from providing answers to a selection problem, to performing movements, or producing changes in the morphology and structure of the machine.

Many of these actuators are studied in the field of robotics, with a large section of the state-of-the-art aiming to emulate in accurate ways human and other biological movements, such as locomotion, hand grasping, etc.

Actuators can be attached with the mechanisms that carry out the sensing, although functionally they are separate. But this is just as a concession to what occurs in nature, as many times in biological entities both sensors and actuators might be one and the same element.

Relating all the above information together, and applying the functional approach, the three main stages of the intelligent process (perception, reasoning and action) would be carried out in machines linearly, using three main elements:

$$\textit{Sensor} \rightarrow \textit{processor} \rightarrow \textit{actuator}$$

This relationship is also fundamental, and it will be applied throughout the book when analyzing intelligent building envelopes for high-performance. It will help to distinguish traditional roofs and facades from our propositions for improved building elements.

Although in practical terms it is not mandatory for intelligent machines to perform all the above stages, a complex system would normally have these three elements. An intelligent system with a high level processor center would also be equipped with feedback systems, in order to evaluate the results of the logic, learn from the outcome and repeat the cycle with modified actions if required, increasing the accuracy of the desired result.

These three stages and their relationship with biological intelligence, allows us to speak of *artificial intelligence*, as a field that deals with the synthetization of intelligent processes to artifacts (Nilsson idem; Poole et al. idem).

Current scientific efforts for the development of artificial intelligence can be found mostly in the fields of computer science and robotics. Perhaps one of their major achievements is the use of logic for deductive reasoning in an inanimate object. Logic is used as a tool for dividing a large problem into smaller steps towards its solution, and which can be translated into programmatic steps.

At the same time, some of the activities that are taking renewed efforts to take artificial intelligence to new levels include improvement of direct knowledge acquisition (such as machine vision, tactile inputs, etc.), and its interpretation into machine-understandable language (Gallo et al. 2015; Sinha et al. 2016). Although still far away, one of the goals of artificial intelligence is to emulate human-level intelligence with all its complexity.

The definition of artificial intelligence denotes that the principles of the process can be applied to objects, also called *agents* in the specialized literature. These agents are emulating the intelligence process, even if they are not thinking as living beings do (which is a major misconception about artificial intelligence).

It can also be understood from the meaning of artificial intelligence, that these artifacts are performing *a metaphor of thought*, which nevertheless can perform more complex functions than machines that do not possess artificial intelligence.

Intelligent agents are autonomous, can be simple or complex, and through the use of knowledge direct their activity towards achieving a goal that is measurable or observable (Russell and Norvig 2013). An agent and the environment in which it acts is called an agent system (Poole and Mackworth 2010). Russell and Norvig divide intelligent agents into five types:

1. *Simple reflex agents*: They act based on a single perception, and provide a single action. They act on well-defined rules of the type *If…Then*. This also allows a straightforward supervision over their functioning, which is usually done by an external agent.

   The environment has to be fully observable or measurable for successful application. Infinite loops are possible and only external actions can help escape from them.
2. *Model-based reflex agents*: Acting on partially observable environments, as they have some types of rules to act concerning the non-observable part. It is a "model of the world", therefore acquiring their name.

The agent keeps track of the perceptions to compare with previous environmental information. It then chooses an action based on the rules, using the same paths as the simple reflex agent.

3. *Goal-based agents*: In addition to the model-based agents, they also have information on the objectives to be achieved. These objectives describe desirable situations.
   The logical system can choose from a series of possibilities, in order to reach one of the objectives. There is some degree of search and planning to prepare the agent to take a given action.
4. *Utility-based agents*: The previous type of agent distinguishes only between a goal and a non-goal state. Utility agents can use a function to measure the degree of desirability of a certain state, given the current condition, actions that can be taken and probable outcome.

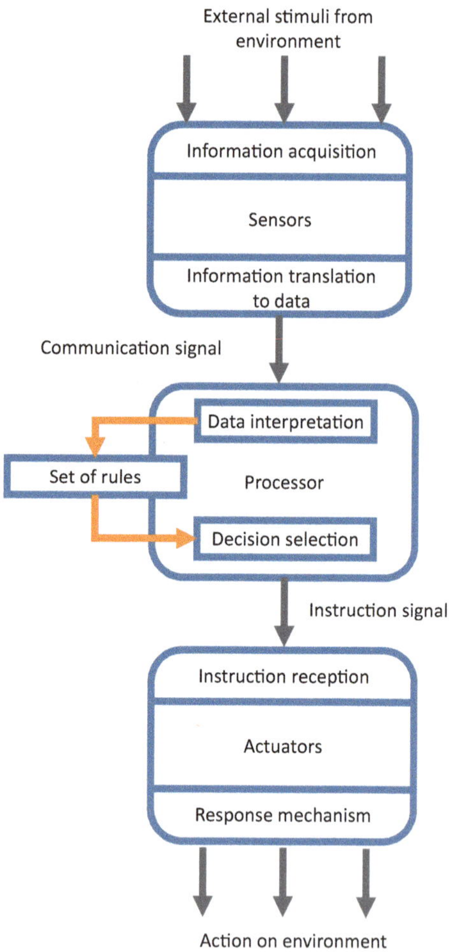

Fig. 1.5 Summary of main elements carrying out the intelligence process for inanimate objects

The agent has to keep record of the environment and model the modifications that can be done on it, giving it some predictive capabilities.

5. *Learning agents*: They can operate in unknown environments and use the outcome of their actions to improve future ones. They can modify their initial knowledge in order to reach a desired and improved performance.
   Theoretically, they have a separate learning element (which makes the improvements to the knowledge base) and a performance element (which takes external actions).
   Improvements are based on feedback given by a "critic", who decides how well the agent is doing. A "problem generator" suggests the actions that can take the agent to perform in new experiences.

Figure 1.5 presents a summary of how the intelligence process is applied to machines. It describes the relationships between sensors, processors and actuators in the same sequence as it was described for perception, reasoning and action. As seen in the text, different types of agents can be derived from this process, helping to conceptualize a series of intelligent envelopes.

1.4 Intelligent and Responsive Architectural Envelopes

If we are to imprint true intelligence in our built environments, then we have to concentrate on providing it to each of their components. The theoretical framework is defined by information presented on the intelligent process and its derivation to artificial intelligence. It can be potentially developed using the different characteristics presented by a variety of intelligent agents.

As this book focuses on providing intelligence to the external layer of the building, from design to specification, it must be pointed out that the building envelope is considered one of the most important elements in the building. It provides an interface between exterior and interior environments, and defines the image we have of a house, office, etc.

Similar to the outer shell of a cell, or the skin of other living beings, building envelopes should be intended to regulate energy interchange with the outside environment, maintaining the general equilibrium for occupants' comfort in the entire space. From an energy point of view, traditional facades account for more than 30 % of the contributions to energy consumption in hot climates (Haase and Amato 2008). In temperate climates a similar percentage can be found, depending on the type of insulation placed (Burton 2015).

However, unlike animal shells and skins, building envelopes at the moment are generally hardly thought as intelligent. This is in part due to their static nature and constructive technologies. Even if they are exposed to changing weather requirements during the year, or the usages inside change rapidly, the façade is perceived as continuing to have the same composition, behavior or image.

As we have seen, a successful intelligent agent is not only a collection of devices and systems that perform complex tasks. It involves carefully including in the process a series of design considerations to fulfill the functions assigned to intelligence when it is applied to inanimate objects such as the building envelope. The expected result from such an intricate process is an object that can minimize operation problems, providing adequate and enhanced responses to different situations that are systematically better than the equivalent "non-smart" counterparts.

According to the theory of intelligent agents, the process can be applied to building envelopes using different degrees of complexity in the arrangement of their parts. As an example, in very simple reflex agent setups, perception sensors are directly coupled to elements that produce a physical reaction. They are demonstrated in the so-called Braitenberg vehicles (Braitenberg 1984), where a wheeled motor is connected to a light sensor. According to the wiring of the sensor, the vehicle is stimulated by light to move towards or away from it. Depending on the amount of freedom available in the wheel arrangement, it can also change course when the initial situation changes. This arrangement is not limited to light, and has been used also for other types of stimuli (Lilienthal and Duckett 2004).

In building envelopes, this type of behavior can be found in some materials and structures, whose properties have been modified to obtain a given characteristic that is different from the usual physical properties found in the original material. They are commonly called "smart" materials (Lazarovich 2012), and the modification can occur at the molecular and composition level.

On the other end, it is not difficult to imagine that the correct assemblage of advanced sensors, processors and actuators can result in more intricate arrangements. An aim to be studied is to project a building envelope to become a learning agent. It could evaluate the outcomes of its decisions, take into account feedback of users as its critics, and improve future responses (based on Nilsson 1998). Figure 1.6 gives a scheme of how the intelligent envelope can be thought as an intelligent agent.

Such examples of possible envelope setup demonstrates that the variety and complexity available to design intelligent systems is large. Architectural designers can be overwhelmed by the universe of options, and some might seem to enclose a given degree of complexity that goes beyond their expertise. In the interest of balancing different project objectives, designers can assign too much or too low a level of intelligence and responsiveness for a given situation. It is also tempting and erroneous to think that smart actuators or processing centers will be the replacement of decisions that have to be made at the design stage.

It is extremely important to make note that designers cannot expect intelligent elements to act smartly, if they are not produced by a proper building design process. *An intelligent envelope is a rational agent; it only does what it is programmed to do.* It cannot be expected to be an omniscient element that will know everything (Russell and Norvig 2013).

The discussion of this book centers on systems complying with the main stages of the intelligent process: perception, reasoning and action. It will give an overview of the principles that are needed to design an intelligent building envelope and the tools available to achieve it.

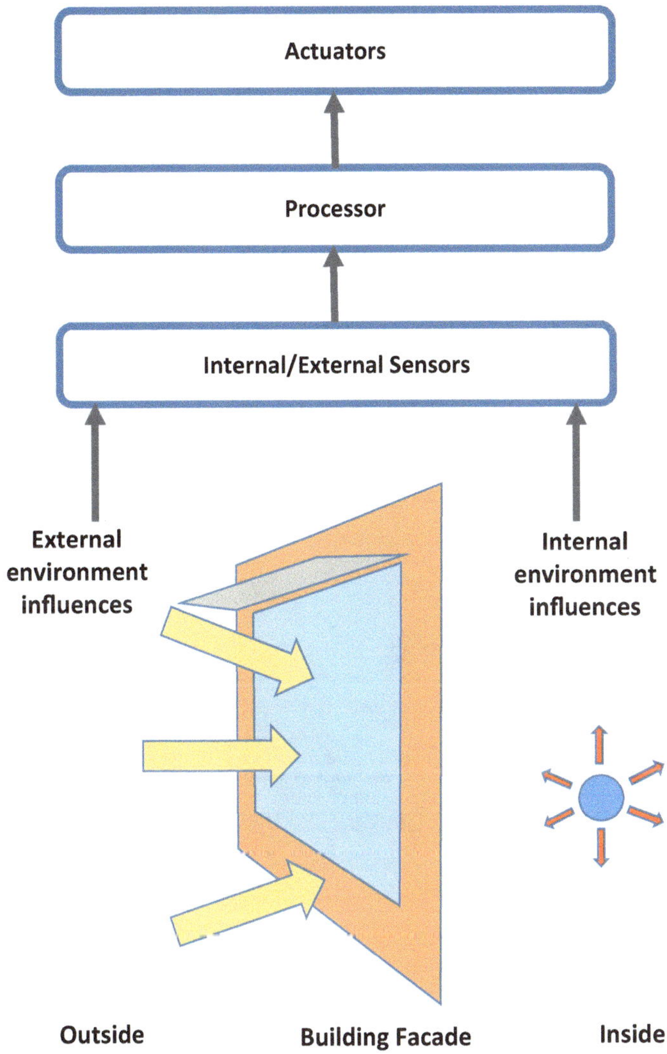

Fig. 1.6 Representation of the intelligent envelope as an intelligent agent

1.5 Types of Inputs and Responses for the Intelligent Envelope

It has been seen that it is possible to make a parallel for the intelligence process found in nature with that being carried out by machines. Nevertheless, the most advanced types of perception and response as performed by biological entities are at the moment not available for all intelligent agents. However, not all inputs and responses need to imitate life in order to act as artificial intelligence.

This section provides an overview of relevant manners of obtaining and displaying information that can be applied in the design of an intelligent building envelope. They are deemed relevant for architectural design in order to help the reader form an idea of what is available. Further information on components will be given in Chap. 2. A sample relationship between the inputs and responses is given by Fig. 1.7.

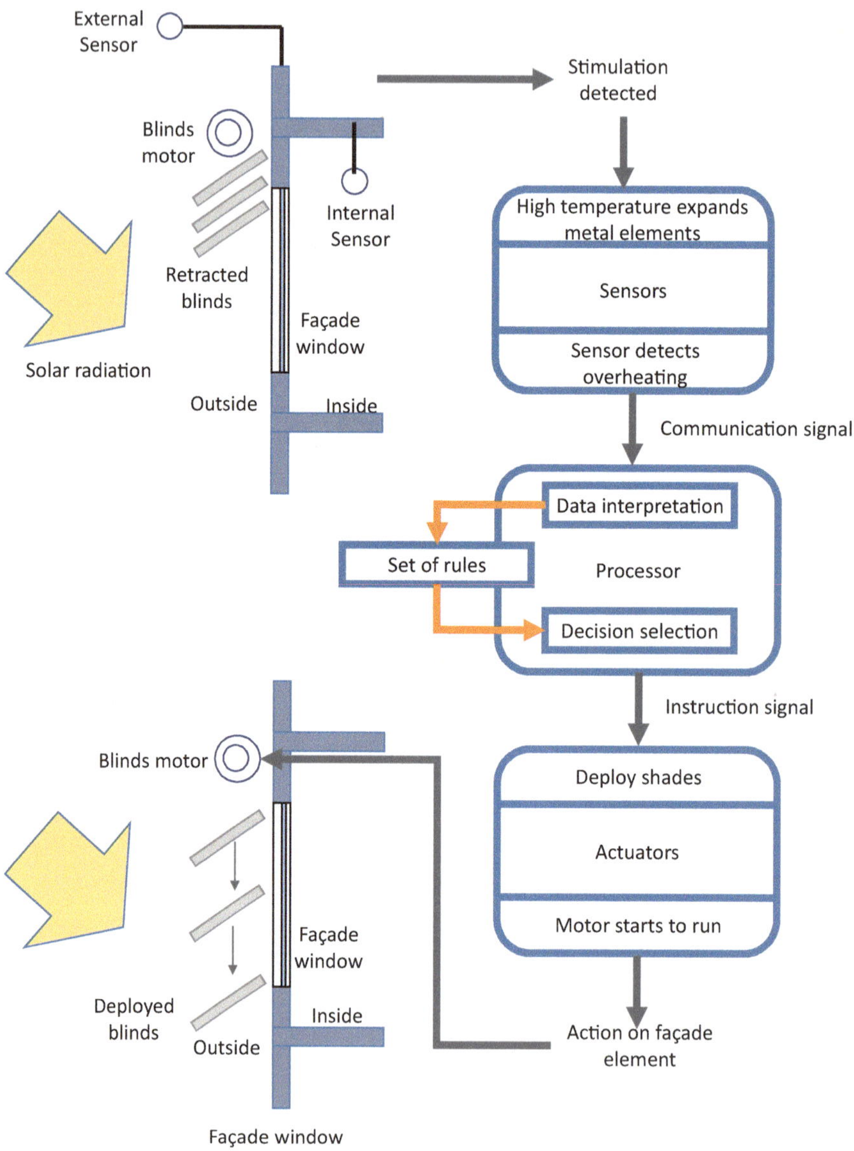

Fig. 1.7 Relationship of inputs and responses for an example of intelligent envelope

Inputs

In the field of artificial intelligence, the environment where perception is carried out can be the real world, or an artificial virtual one such as the Internet (Poole and Mackworth idem). The environment can be partially or fully observable, in which measurements can be done on different variables, relevant for the intelligent agent operation.

Nevertheless, for architecture, virtual environments are more the realm of simulation, in order to predict the performance of the envelope. Real environments would be exemplified by the building being commissioned and occupied, subject to the changes in weather, time and usage.

Perception, as the collection of knowledge from the world, will become a signal representing information. It will be abstracted for interpretation by the processing center. It is also important to decide which information has to be collected and which can be made note of, as not all of it has the same weight when considering overall design. Further reading on this subject is found in Chap. 5.

An examination of relevant literature offers a wide variety of methods to classify inputs taken by intelligent systems. However, in order to focus on what is applicable to intelligent building envelopes, the following classification is proposed, adapted from Russell and Norvig (idem):

(a) *Input from program interfaces:* Data is simplified to electrical signals that represent numerical terms. They are entered to the system in common ways such as turning up or down knobs or buttons, or by writing numerical data using a computer screen.
(b) *Input from the climatic environment:* A sensor or an array of them is placed strategically in order to passively gather information from the environment, such as temperature, solar radiation and relative humidity. The electrical impulses sent by the sensors are translated to data understandable by the processing center.
(c) *Input from actions in order to obtain information:* The system could gather information actively in order to "know" what is going on inside or outside the internal environment. For example, at given intervals it could activate an occupancy sensor to determine the number of occupants present. It would then decide if lights can be turned off and do nothing, or if blinds should be deployed or raised to regulate incoming solar radiation and light conditions.
(d) *Input from human language:* A very uncommon option at the moment, users could interact vocally with the system simulating a conversation, prompting the system to act according to what is gathered. The most successful systems at the moment only operate on simple voice commands such as those given to smartphones to start some actions. Technical difficulties arise when interpreting simple sentences without any other information available. The need for more information (and consequent dialogue to obtain it) could frustrate users or make them provide wrong information.

Designers would need to consider which of these inputs is more relevant for each particular case, but the list shows that these possibilities exist.

Responses

Equivalent to the environment where inputs are recorded, responses can be real or virtual. Virtual responses are found commonly in the output provided by computer programs, or when performing simulations previous to the commissioning of a system.

Of more interest for our subject are those responses that produce a change to the intelligent agent or its immediate environment. When applied to the building envelope, real responses can be active, passive or a mixture of them.

(a) Active responses produce an interaction where the actuator has a noticeable change from its initial situation. Movements and their directions can be described by the degrees of freedom present in the mechanism.
Usually many elements are involved in producing this change. One of the most commonly found examples involves roller shades or blinds going up or down to block solar radiation. In this case there is only one degree of freedom.
(b) Passive responses are created by pure physical or chemical reactions that result from stimulus to the material itself. Smart materials also fall in this category (Janocha idem, Lazarovich 2012), as the reaction occurs down to the molecular level. Specific properties of materials are modified or taken advantage in such way that their molecules can be considered micro or even nano actuators. An example can be found in molecules that can be manipulated to selectively block certain wavelengths in an electrochromic window (Llordes et al. 2013). However, there is no change in the overall direction of the component that houses them (the smart window).
(c) A combined response of both active mechanisms and passive materials is possible. For example, a physical reaction such as the expansion of a shape memory alloy can trigger a large-scale movement that can be noticed by observers (Lazarovich idem).

A sample relationship between the inputs and responses is given by Fig. 1.7, showing the case for control of solar radiation as a trigger for a response using blinds to control it within the intelligent façade. The process will continue with internal sensors detecting any reduction in internal illuminance levels on certain areas of the building, which can trigger supplementary lighting, and so on.

The potential within each division of both inputs and responses is quite large and enables different possibilities for creativity in architectural design as well as achieving true energy saving and improved comfort conditions. The chapter has provided the foundations for understanding the familiar outer layers of a building as envelopes for a system that has to become intelligent for improving its performance and offer better comfort conditions to occupants.

This book will provide an insight for application of the artificial intelligence theory. On Chap. 2 we present built examples that have applied these principles, offering a background for studying the implementation of intelligent building envelopes. The remaining chapters will detail the tools and framework for design.

References

AHD American Heritage Dictionary (2016) Machine. www.ahdictionary.com. Accessed May 2016

Bach J (2009) Principles of synthetic intelligence—PSI: an architecture of motivated cognition. Oxford University Press, New York

Braitenberg V (1984) Vehicles: experiments in synthetic psychology. MIT Press, Cambridge

Burton S (ed) (2015) Sustainable retrofitting of commercial buildings: cool climates. Routledge, Abingdon and New York

Corsini RJ (ed) (1987) Concise encyclopedia of psychology. Wiley, New York

Encyclopedia Britannica Online (2015) Life. http://www.britannica.com/topic/life. Accessed Dec 2015

Fagin R, Moses Y, Vardi MY, Halpern JY (2004) Reasoning about knowledge. MIT Press, Cambridge, Massachussets

Gallo S, Son C, Lee HJ, Bleuler H, Cho I-J (2015) A flexible multimodal tactile display for delivering shape and material information. Sens Actuators A 236:180–189

Goel V (2007) Cognitive neuroscience of thinking. In: Handbook of neuroscience for the behavioral sciences. Wiley, New York

Haase M, Amato A (2008) Sustainable façade design for zero-energy building in the tropics. In: Clever Design, Affordable Comfort—Proceedings of the 23rd international conference on passive and low energy architecture (PLEA), Geneva, Switzerland, 6–8 Sept 2008

Hayes-Roth B (1995) An architecture for adaptive intelligent systems. Artif Intell 72:329–365

Janocha H (ed) (2007) Adaptronics and smart structures: basics, materials, design, and applications. Springer, Berlin

Lazarovich N (2012) Smart material systems for high performance building envelopes: guidelines for smart material implementation in the design of climatically responsive building envelopes. Dissertation, Technion Israel Institute of Technology

Lilienthal A, Duckett T (2004) Experimental analysis of gas-sensitive Braitenberg vehicles. Adv Robot 18:817–834

Llordes A, Garcia G, Gazquez J, Milliron D (2013) Tunable near-infrared and visible-light transmittance in nanocrystal-in-glass composites. Nature 500:323–326

Merriam-Webster (2015) Intelligence. www.m-w.com. Accessed Dec 2015

Nilsson N (1998) Artificial intelligence: a new synthesis. Morgan Kaufmann, Burlington

O'Brien DP (2009) Human reasoning includes a mental logic. Behav Brain Sci 32:96–97

Oxford Dictionaries (2015) Intelligence. www.oxforddictionaries.com/. Accessed Dec 2015

Poole D, Mackworth A (2010) Artificial intelligence: foundations of computational agents. Cambridge University Press, New York

Poole D, Mackworth A, Goebel R (1998) Computational intelligence: a logical approach. Oxford University Press, New York

Reber AS (ed) (1985) The Penguin dictionary of philosophy. Penguin, London

Rowland M (1992) Biology. Thomas Nelson and Sons Ltd, Surrey

Russell SJ, Norvig P (2013) Artificial intelligence, a modern approach. Pearson, Essex
Schacter DL, Gilbert DT, Wegner DM (2011) Psychology, 2nd edn. Worth, New York
Sinha Y, Jain P, Kasliwal N (2016) Comparative study of preprocessing and classification methods in character recognition of natural scene images. In: Machine intelligence and signal processing. Springer, India
Tortora GJ, Derrickson BH (2014) Introduction to the human body, 10th edn. Wiley, New York

Chapter 2
History and State of the Art

2.1 Early Developments and Postulates

Technological developments leading to the current state of the art for intelligent building envelopes have a long history. The design and building activities move at a particular pace, and the efficacy of new technologies can only be proven with particular examples. The following account is not exhaustive and pretends to highlight only some representative works of the many examples available.

The first dwellings crafted and used by human beings had the obvious intention of creating spaces that would provide protection and comfort from the external environment. But changing conditions during the day and the year would have required some type of internal environment regulation to avoid extremes during day or night, winter or summer.

Even though concrete evidence of building envelope regulation in prehistoric times is scarce, some early tribes had dwellings made out of organic materials such as the wigwam. In it both structure and envelope can be moved to allow ventilation and privacy as needed (Canadian Encyclopaedia 2016), constituting a primal example of an adaptable cover.

Early examples of ancestral Native American pueblos in New Mexico, Arizona and Colorado, show adaptation to changing conditions during seasons and hours of the day, making intelligent use of the geometry and orientation of the overhanging cliffs where the buildings were located, and the correct use of the heavy thermal mass of the stone.

In traditional or historic construction, internal regulation depends largely on the thermal properties of static materials composing the opaque area of the enclosing walls. This is experienced in well-made ancient buildings made of stone, mud, wood or even those insulated with cellulose or straw (Bozsaky 2010), of which abundant examples exist such as ancient indigenous adobe buildings in the Americas (Keoke and Porterfield 2003). Nevertheless, the fragility of such materials requires constant and dedicated preservation measures (Technical Preservation Services 2016).

G. Capeluto and C.E. Ochoa, *Intelligent Envelopes for High-Performance Buildings*, Green Energy and Technology, DOI 10.1007/978-3-319-39255-4_2

Historically there are also many surviving examples of adaptable openings, which would become the inspiration behind contemporary devices. These include old "box-type" windows, commonly found in Eastern Europe, that have double windows to protect against cold weather. They do work as a small-scale double facades (Fig. 2.1). Also to be mentioned as historical examples, are the well-known movable wooden shutters, which provide privacy, night insulation when needed, protect against excessive heat and light penetration in Mediterranean and tropical climates (Fig. 2.2). However, operation of these traditional elements still depends on manual actions by the user.

At the turn of the 20th century, many building types had already become more complex than their traditional precedents, new usages were required, and novel engineering technologies were becoming available. Architects understood that the design of building components such as the external envelope could not strictly follow traditional design methods due to the requirements of modern buildings. Facades and roofs required an organized planning process directed to the accomplishment of formal and functional objectives that could have a more tangible evaluation.

One early example of modern detailed envelope design corresponds to the Paimio sanatorium, planned by Alvar Aalto in 1932. The envelope layout was produced after incorporating relevant medical professionals in the design team, which for the time was a revolutionary way of working.

Fig. 2.1 Traditional double windows in Warsaw, Poland. *Photo credit* Carlos E. Ochoa

Fig. 2.2 Traditional wooden shutters in Rome, Italy. *Photo credit* Carlos E. Ochoa

Careful detailing was placed on the window as a generator of health. Its components had enough parts to manually regulate air intake and shading as desired by the patients, therefore giving the possibility of changing the micro climate inside the room. Architectural elements such as sun-bathing terraces were also placed according to their most suitable orientation (Anderson 2010; National Board of Antiquities 2005).

Le Corbusier, on the other hand, also introduced the concept of *brise soleil* and detailed facade sub-division. Such concepts were based on observations he made of traditional window latticework during his travels to Mediterranean countries. He also tried to include climatic principles in the façade designs, with views of some sort of adaptability to local climates.

Although now we can evaluate his work in retrospective, the execution of his climatic principles was not always successful. For example, some of the methodologies he used tended towards the empirical, and throughout his *ouvre* there were few exceptions to his generalized utilization of materials such as concrete and glass despite the location.

However, it was notable that, based on earlier industrial achievements for mass production, the architect had among his many propositions that buildings would be used as "machines" for living inside them. Such machines were to be covered with "breathing skins" (Curtis 1994).

2.2 Historical Points of View for Intelligent Building Envelopes

2.2.1 The Highly Mechanized Envelope

Such view of a self-regulating, highly mechanized building, would have its first truly operational interpretations during the last decades of the 20th century. This period also coincides with breakthroughs in the development and generalization of sensing capabilities and electronic technologies, such as the personal computer and the microprocessor, which before had been very expensive and unpractical products. The introduction of these elements to a wider audience, and for a larger variety of applications, allowed the beginning of a significant size reduction process for electronic elements. As a result, portability and applicability of computers and microprocessors in new situations, such as building components and controls was considered.

In 1980, the nine-story *Occidental Chemical Center* in Niagara Falls, New York, was completed. The building featured many technological advances for the time: Thermal insulation was provided by a double façade with outer insulating floor-to-ceiling glazing, ventilation was controlled by electronic dampers, and daylight was regulated by inner automated aluminum louvers. Quite advanced for its time, it was hailed as the future of building and of low energy consumption.

However, over-dependence on mechanical elements, and lack of control of users over their working environment, made it sadly become an example of what can go wrong with automated buildings. The systems were too complex for that period, and never received adequate maintenance, leading to their malfunction. System design was made according to a usage model that was supposed to be stable during the years, and which proved otherwise. In addition, external and unforeseen factors occurred, such as changes in occupancy and ownership. Even the decision of the new owners to change the landscaping to a parking lot, and the construction of neighboring buildings, meant that the systems could not work as intended. Eventually the components were replaced by extremely cheap substitutes. Nowadays the building is almost empty, in need of repairs and its space is destined to other uses different than the original ones (Fortmeyer and Linn 2013).

In 1983, *City Place I* in Hartford, Connecticut opened to the public. The building comprises 38 stories and was designed by Skidmore, Owings and Merrill (SOM). Although today it would be defined as a "highly wired building", it was planned to coordinate HVAC, lighting, security, communications and other systems through a central computer network with the stated aim of making energy consumption more efficient. Its data was gathered from sensors rather than following a program, and also included different user scenarios (New York Times 1983).

This made it one of the first buildings of its kind that incorporated computer systems and interactive internal climate control. Although it used a more conservative approach and depended on mechanical systems (a well-known strategy available during the 80s), the building is still in use today.

Some years later, in 1987, Jean Nouvel's *Arab World Institute in Paris* was finalized. One of its distinguishing features is the use of 240 photo-sensitive modules with 30,000 motorized diaphragm apertures placed on different facades behind glazed surfaces of the main building (Figs. 2.3 and 2.4). Fabricated in metal, the diaphragms open and close according to conditions determined by light sensors.

The intention by the designers was to ornament the façade and reduce the amount of light entering the space, using a modernized version of the *mashrabiyas*, a window lattice from Arab countries (Hanser 2006) that allows achieving privacy as well as shading and ventilation passively, making good use of the geometrical proportion of its components. Although the devices create a special aesthetic quality, the diaphragms depending exclusively on complex mechanical systems encountered technical problems, and their maintenance proved complex and expensive for the building owner (Fortmeyer and Linn idem).

The *GSW headquarters* in Berlin, was designed by Sauerbruch Hutton and completed in 1999. One of its particular characteristics is a series of colorful shades, in red and pink hues placed on the West façade of the new tower (Fig. 2.5). The shades slide sideways, regulating solar radiation, while being located inside a double façade with a cross-ventilation system (Fig. 2.6).

Climate strategies (such as spray cooling) are managed by a central system which gathers information through sensors. However, in contrast to the fully automated approach, a range of aspects such as ventilation, shading and lighting can be regulated by users. This creates a changing pattern in the distribution of openings in the canvases. The tower was intended as a generator for reconstruction in the reunited city. It also provides it with a noticeable visual urban reference, through the particular coloring scheme of its dynamic canvases (Wigginton and Harris 2002).

Fig. 2.3 Façade of the Arab World Institute Paris with the operable diaphragms. *Photo credit* Carlos E. Ochoa

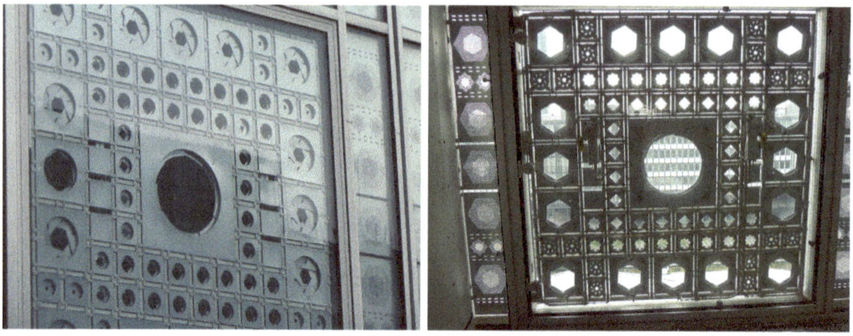

Fig. 2.4 External (*left*) and internal (*right*) details of the operable diaphragms of the Arab World Institute Paris. *Photo credit* Carlos E. Ochoa

Fig. 2.5 West façade GSW building, Berlin. *Photo credit* Guedi Capeluto

Fig. 2.6 Detail GSW building shades. *Photo credit* Guedi Capeluto

2.2.2 The Naturalistic Envelope

The decade of the 1990s saw a more widespread manifestation of ecological and energy saving ideals as design generators in architecture as part of a new mainstream school of thought, where nature is an important factor to be considered. Climatic design principles had reached maturity, and started to be taught as part of an increased number of architectural education curriculums.

A post-modernism point of view was introduced to architecture in a more widespread fashion. Elements used in traditional architecture were re-evaluated for their contributions to comfort. They were perceived as intelligent solutions executed in a time when the only choice was the lowest or no energy use. Traditional construction technologies that use thermally comfortable materials, such as adobe, were reevaluated through a more scientific scope towards their use.

New construction technologies, developed with the specific objective of minimizing energy consumption, were developed and brought into practice. It was also acknowledged that mechanical systems could not become the only source for occupant comfort. Many of these ideas were implemented into architectural designs around the world as these ideas found wide acceptance.

In contrast to the highly mechanized approach under this view, building envelope systems were envisioned not as separate entities that are solved later on by mechanical engineers, but as integral components that have received detailed thought from the start of the design process. In addition, calculation methods for energy and performance prediction became more accessible through newly developed computer programs. They allowed parametric exploration of different design options and more accurate prediction of the expected performance.

An outstanding example of this new vision for intelligent building envelopes is found in 1993s *Consorcio Vida building* in Santiago, Chile, designed by Enrique Browne and Borja Huidobro. One of the main distinguishing features of the project is the use of dynamic shading provided by deciduous vines that grow to almost the full size of the West façade. They were conceived as a "vertical garden", sixteen stories high (Fig. 2.7) adding a green area that contributes to improve the environmental quality of the zone.

In contrast to conventional buildings, where their image deteriorates during the years, this solution provides the building with an always changing aspect with permanent validity. As Browne states: "architecture and vegetation are fused together, turning the latter into a construction material, keeping the building relevant for an undefined time" (Arqa 2011).

The color and density of the vines change according to the seasons, regulating solar light that comes into the office spaces. This in turn also helps to save energy on heating, cooling and lighting, providing a better working place. They also create a moist microclimate in the dry Santiago valley where it is located (Fig. 2.8). Vegetation is also used in the lower floors in the form of full-grown trees to provide shade to pedestrians and protect the façade of lower floors (Senott 2004).

Fig. 2.7 Consorcio Vida building, Santiago, Chile. Winter (*up*) and summer (*down*). *Photo credit* Guedi Capeluto

Fig. 2.8 Consorcio Vida building, Santiago, Chile. Details of the trellis with vegetation in the flowering stage. *Photo credit* Guedi Capeluto

2.3 Recent Developments and Examples

2.3.1 The Integrative View

The most recent generation of intelligent envelopes combines the availability of new active technologies, such as smart materials and robotics, with a wider flexibility in the application of passive strategies. It also makes use of expanded and improved computational capabilities, for implementing more reliable electronic envelope elements. It also uses this new computing power for routine energy analysis of multiple options in shorter time, and through more user-friendly interfaces.

Such advances allow intelligent envelopes the possibility to reach higher levels in the list of intelligent agents presented in Chap. 1. Computing advances allow artificial intelligence principles to be programmed in the processing areas. The entire composition of the building envelope starts to emulate more the principles of a living organism.

These improvements are also required by necessity: low energy consumption and ecological principles have become part of the mainstream in many countries, and are no longer the ideas of visionaries. Building codes and directives now mandate maximal energy usages that cannot be trespassed, they also oblige designers the incorporation of certain energy saving measures, as well as specific performance evaluation methods.

The possibilities to plan intelligent envelopes have increased with the integration of passive and active technologies, due to the increase of reliability in electronic devices. Intelligence in a building is no longer a one-off product that has to be started from scratch. Currently there are commercially available protocols such as KNX, LonTalk, ZigBee, DALI, etc. that make computer-based building envelope management to be standardized and relatively accessible in terms of price and operation. Advanced sensor devices providing better integration of different devices within the envelope and with other parts of the building (KNX 2016) by monitoring also the physical capabilities of passive elements.

Novel material technologies have also gained ground in recent times although still require wider market diffusion. A wide range of the so called "smart materials", since they "know" how to behave in a regulated way according to each situation, still need to be thoroughly tested through experiments before being applied on the field. Nevertheless, it is a field of research that promises new and groundbreaking directions that will change the way that we think about construction.

Some examples that can be mentioned, of intelligent envelopes created using these new principles, include the 18-story *San Francisco Federal Building*, completed in 2007 and designed by Morphosis. HVAC systems are present only in lower floors with high occupancy and equipment concentration. It also pushes the concept of outer envelope in a literal way, through the use of a folded metal sun shading pattern which also becomes the signature aesthetic of the project (Arcspace 2007) (Fig. 2.9).

Fig. 2.9 San Francisco Federal Building by Morphosis, South-East facade. *Photo credit* Guedi Capeluto

The building relies on a series of smart controls, directed through a computerized control system, to enforce a series of climatic strategies such as cross ventilation, night ventilation and maintenance of lighting levels. The interior planning maximizes also airflow and daylight penetration, together with direct outside views for all the workers.

On the other side of the Atlantic, at a smaller scale, the *Oscar von Miller Forum* in Munich was designed by Thomas Herzog and completed in 2009. It is a mixed-use project that combines offices, learning spaces and temporary housing. The building is also a good example where different passive and active strategies were combined in order to minimize energy expenditures. Examples include automated control of natural ventilation and shading with overrides, district heating systems whose flows take into account the amount of heat obtained by passive thermal mass, domestic hot water collectors, use of roof gardens, etc. There is also some self-energy generation through the use of photovoltaic panels (Fig. 2.10).

These systems are managed by different computer control centers that "know" how to combine energy gained from both active and passive systems. The control centers are also programmed to combine the strategies used by one system for the benefit of other systems (Herzog 2010).

The *KfW Westarkade* high rise in Frankfurt, completed in 2010, was also designed by Sauerbruch Hutton. One of its main peculiarities is having a very low energy consumption for its built area and height. Following the recent line of thought about intelligent envelopes, a series of passive and active features are

Fig. 2.10 Shades, photovoltaic panels and hot water collectors Oscar von Miller Forum, Munich. *Photo credit* Carlos E. Ochoa

combined. One of its distinguishing features is the so called "pressure ring". It is made of a double façade that takes advantage of prevailing wind directions and has the serrated shape that characterizes the project. This allows natural ventilation through operable internal windows, neutralizing any turbulence effects associated with high towers. In turn, the building does not require being sealed and natural ventilation can be used even in the highest floors.

The openings and external flaps allowing air inside are computer controlled according to the internal requirements and prevailing wind conditions. Sunlight redirection systems allow savings in artificial lighting, and a better use of the plan distribution. The main heating sources are geothermal and activated slabs, but any excess heat generated by loads such as machinery or computer equipment is also reused (CTBUH 2011).

We have presented these well-known available examples, to prove that a wide variety of possibilities for intelligent envelopes exist. They have been shown as the result of different design philosophies, which affect the final arrangement and dependence between systems. Historical developments have also shown what can be learned from over-dependence on a single mechanical or electronic system, which is not recommended. A more integrative approach is preferred, since it can take advantage of the benefits from both passive and active systems. It also provides an enriched design with high intelligence potential.

2.4 Classification System for Intelligent Envelope Components

Trying to understand the large variety of envelope components is complex. To do so, a systematic approach is required. The use of such approach is to take advantage of what is available nowadays, and to have a hint of what will be available in the very near future. A classification system is presented in this section, with the aim to organize the component inventory and motivate possible new research.

As it has been seen, there are currently numerous options and approaches to develop an intelligent envelope. An overview is offered in this section, where elements are classified to understand how components work. By this, readers can clarify and organize design efforts, exploring the development of new elements where few of them exist. The classification system intends to be indicative, as there are always innovations in the fields of new materials and management systems. Categories and divisions in the classification system also follow the separation made of intelligent elements into sensors, processors and actuators. Different sub-categories are also noted.

2.4.1 Class A—Perception/Input Elements

Elements belonging to this class gather information about the internal and external environment, and also monitor any changes in them. They can also receive signals from users that are interpreted as commands for an action to be taken. Both feeds constitute the input to the intelligent envelope. The main sub-categories in this class are:

(a) *Sensors*: They can be placed inside and outside the envelope, and can measure one or many environmental aspects at the same time. Some of these include: internal and external temperature (dry/wet bulb), external and internal relative humidity, wind speed, illuminance over a given surface, human presence or absence, electrical appliance usage, etc. Although composed of artificial materials, and only dedicated to collect information, they have been compared to the nerves of a living being due to the information that they can sense from the world (Giurgiutiu 2007).
Their size is much reduced and can be almost undetected by casual passers-by. They work by conductance, resistance or sensitivity. For example, a photocell is composed of semiconductors that absorb photons from a light source, modifying the resistance range (Diffenderfer 2005).
Sensors can be connected to the processing centers either as part of a wired network or through wireless communication (Towler 2011). This allows their strategic placement in different locations, thus having to endure different environmental conditions or connection requirements.

(b) *User interfaces*: They can vary in complexity from classical on/off switches, thermostats, to remote controls and touch-screen panels. Although one action is performed by the user on the interface, the control themselves can interpret such action as the instruction to regulate different activities at the same time. Different degrees of sophistication exist in the way that orders are entered and the information is presented to the user (for example, electronic displays, turn knobs, etc.). Nevertheless, the area of user interfaces requires further developments in order to exploit the full potential that intelligent envelopes offer to end-users, facility maintenance, developers and architects. Their usage can also be tied to different conditions unrelated to performance, such as budget available to use touch screens instead of switches; also to the type of project, as it might add further layers of complexity that are not desired such as adding a computer interface for a room that is seldom used, etc.
The implementation of now-familiar interfaces for electronic devices such as smartphones or tablets still has to find widespread application in the controls used within intelligent buildings (Henson 2012).

2.4.2 Class B—Control Processing Elements

In this class we can find the components that gather information collected in the input phase through sensors and interfaces, and "decide" on the course of action to take. These decisions are based on different logical steps, pre-defined policies to be carried out by actuators and that were decided by a design team. The orders, translated as electrical impulses, are then sent to the relevant parts. For this class we shall discuss the key policies that must be taken into account by any intelligent envelope protocol, and the main types of control systems available.

Policies must be applied when specific events occur outside the envelope or within its enclosure. Designers have to consider beforehand the desired actions that actuators will take according to limits imposed by changing weather and occupancy events. These policies, expressed as strategies, will then be translated to specific commands. As seen in Chap. 1, there are different ways to express the reasoning for an intelligent agent. Some include only reactions to the environments as the most basic unit. Using logic systems might comprise a sensing of the internal and external environment, to decide on the best policy to implement. More advanced logic systems can include a simulation being done, with evaluation of the possible alternatives. Nevertheless, at the moment it is more common to encounter reactive elements, while the programming complexities of logic systems limit their wider application.

As an application example to control overheating, when sensors detect that direct solar radiation goes beyond a certain threshold, shading elements should be deployed potentially to a given angle that blocks solar radiation. This is a reactive case.

As a second example, the user can enter directly the number of occupants that will be present in a space for a given period. Therefore, the system can then simulate and calculate the amount of time that openings must function in order to provide additional ventilation so that the expected number of users will have enough quality ventilation, and keep this result in case the processing center did not receive any feedbacks or modifications during the planned occupancy time. This is an example of a learning system.

The main policies that can be mentioned are the following:

(a) *Lighting*: How illuminating elements will be used to fulfil daylight supplement, task visual performance requirements, controlling glare, etc.
(b) *Shading*: Operation types of internal or external shading devices to avoid overheating and glare according to their type (e.g. shades, blinds, etc.), activation limits and modes such as angle variations of different blind sectors.
(c) *Thermal comfort*: Regulation of temperature and relative humidity for human comfort according to the number of occupants and activities performed, as well as the needed actions to protect equipment and furniture from climate extremes when the building is not occupied.
(d) *Ventilation*: Strategies to promote or cancel the entrance and movement of air in a space. Air movement can be within the same area, or there can be an introduction of air currents inside or outside the space.
(e) *Energy management*: This strategy relates to how energy is administered in terms of consumption, and in certain cases, for energy self-generation. It also concerns the usage of energy from certain envelope elements that use electric energy in order to keep a balance.
(f) *Administrative policies*: It is referred here to envelope operation in case of unusual events (such as, for example, extremely high winds), maintenance, diagnostic, power failure, emergencies and security (for example, giving an order for blinds to be deployed during night time in order to prevent break-ins). They are necessary in order for day-to-day functions to be carried out, even if they are not directly related to energy or climate.

Processing elements are the ones in charge of administering the policies according to the conditions pre-set by designers and imposed by end users. They carry out instructions to perform a series of actions that will result in the regulation of the envelope or the internal space. These elements can be programmed to have diverse degrees of adaptation to changing conditions. There is also a large variety in the type of logic that can be applied, and the order of strategies or their combination that need to be implemented through actuators.

The following options to manage the policies are available, listed in decreasing order of electronic or mechanical complexity. They also show how different selections for control centers are available according to project constraints such as available budget, or maintenance scheme:

(1) *Building management system* (*BMS*): Almost every aspect of the building is supervised by a central computer decision center. This includes acclimatization, mechanical systems and building administration. The most advanced models incorporate current advances in artificial intelligence for computers. However, in order to avoid general failures, BMS are composed of modular elements where each of them is dedicated to a given function. Thanks to their complexity, a number of scenarios can be programmed for different types of responses. This setup is suitable for buildings that are administrated or maintained in a centralized fashion. A BMS is the most commonly found type of management system in intelligent buildings and envelopes.

(2) *Synchronized controls*: In this arrangement, there is no central decision center, and all policies are administered by different management modules. Sensors, processors and actuators are grouped to perform one activity and have one specific goal. Coordination is necessary between the different sub-systems. An advantage is that failure of any system is contained to one sector, and does not affect the operation of the entire setup. One of the main disadvantages is the elevated number of sensors and modular control processors needed, and that their maintenance might take longer. However, it is highly convenient for buildings with different and changing tenants. One example of these type of controls was implemented in the New York Times building (LBNL 2016a). An important variation of this system occurs when there is no processing center, and actions are performed as direct physical or chemical response to the environment. Many smart materials operate under this principle. Examples include independent shape memory alloys, electrochromic windows or dimming elements connected to an array of light sensors, among others.

(3) *Modular management*: as a hybrid between BMS and synchronized controls, entire prefabricated envelope modules are constructed. Each of the modules contains their own management system for lighting, ventilation and shading. An example is the conceptual TEmotion façade (Wicona 2016).

(4) *Schedule programming*: These correspond to fixed, cyclical operation modes. There is no immediate response to short-term environmental changes, but they are adequate for long term operation of certain systems, such as ventilation or HVAC. They are performed by very low cost components such as thermostats or timers, and their operation is almost intuitive.

(5) *Passive building*: Although there is no control system involved, strategies are administered by the physical properties of static elements in the envelope, which cover the policies mentioned previously. In this case, intelligence resides in careful architectural design and material selection, using well-studied bioclimatic design principles that are appropriate for each climate. They can co-exist with highly sophisticated electronic building systems (Fig. 2.11).

Fig. 2.11 A passive shading system as second skin for a building and signature design feature. San Francisco Federal Building, designed by Morphosis. *Photo credit* Guedi Capeluto

(6) *Users operation*: Direct human interaction occurs with building components, which are operated to restore visual or thermal equilibrium. For acceptable results, good design provides with different mechanisms for fine-tuning this operation. Examples are segmented window openings, division of lighting arrays into different sectors, operable shading devices and ventilation flaps. Behavior patterns can be abstracted and predicted from research and probability theories (Reinhart and Voss 2003).

2.4.3 Class C—Actuator Elements

This class contains a large and always increasing number of elements, and the most visible in the intelligent envelope. Actuators usually perform one function in order to execute the objectives set out in the policies. Elements themselves can be conventional but operate under dynamic modes, or they can be innovative and depend on material properties or specific technical developments, such as "smart materials". The main system groups that can be currently found are the following:

(a) *Daylight and shading management*: They regulate direct and diffuse solar radiation that enters a space and can help with privacy (Fig. 2.12). The most advanced models provide solar tracking capabilities in order to maximize daylight utilization. Some of them can also provide movable insulation. They can be subdivided into sun shading elements (shades, blinds, louvers); daylight redirection elements (lightshelves, reflective blinds); and daylight transport elements (fibre optic pipes, etc.).

Fig. 2.12 Solar radiation reactions from the façade to a cloudy day (*left*) and a sunny day (*right*). Oscar von Miller Forum, Munich. *Photo credit* Carlos E. Ochoa

(b) *Fenestration*: Elements are related to the openings in the envelope (windows and clerestories). Glazing constitutes the majority of components found in this division, and can range from conventional panes to specialized innovative applications such as insulating, light redirecting or electrochromic changing glass (Compagno 2002; LBNL 2006).
(c) *Cooling, heating and ventilation*: Elements in this group help to mechanically or naturally achieve a given air temperature while reducing overall energy consumption when compared to conventional systems. The main subdivisions constitute passive and active systems. In passive systems materials are used to store or release heat. New developments include super-insulating materials with minimal thickness, materials that provide high thermal absorption and adequate time-lag heat release (Jelle et al. 2010). For active systems, the overall U-value of the envelope is varied according to the climatic needs of the season, by increasing ventilation rates in a double cavity in order to generate additional heat, or by circulating fluids inside a tube system that becomes heated by the sun (Fig. 2.13). Examples include water- or solar-based cooling, concepts in opaque movable insulation, or solar vented cavities.
(d) *Energy generation*: This section includes the different options available to generate electricity, ranging from typical grid connection to self-generation such as envelope photovoltaics, wind turbines, and even conceptual panels that

Fig. 2.13 Solar thermal collectors on the roof as supplemental heating system. Oscar von Miller Forum, Munich. *Photo credit* Carlos E. Ochoa

generate biomass for a building (Arup 2016). Some of these technologies interweave with other systems such as fenestration and daylight (Fig. 2.14). The total output of generated energy has been increasing in recent years compared to earlier developments.

Fig. 2.14 Solar photovoltaic panels used also as shading devices. Oscar von Miller Forum, Munich. *Photo credit* Carlos E. Ochoa

Table 2.1 Summary of the classification system for intelligent envelopes

Class	Category	Sub-category
A. Sensor/input elements	Sensors	Internal and external environmental aspects
	User interfaces	One-function buttons and switches
		Remote controls
		Computer monitors and touch screens
B. Control processing elements	Policies	Lighting
		Shading
		Thermal comfort
		Ventilation
		Energy management
		Administration
	Policy management	Building management system
		Synchronized control
		Modular management
		Schedule programming
		Passive building
		User operation
C. Actuating elements	Daylight and shading management	Sun shading
		Daylight redirection
		Daylight transport
	Fenestration	Conventional glazing systems
		Innovative glazing systems
	Cooling heating and ventilation	Active options
		Passive options
	Energy generation	Grid connection
		Self-generation (PV, wind, etc.)

A summary of the classification system is presented in Table 2.1. It presents the groups where different technologies can be placed. Every day more examples of each category are being developed. This table also demonstrates the large potential of element combinations that can be made available.

2.5 Comparison of Two Smart Actuator Systems

This section describes how the existence of different technology levels in actuators can be used in different design levels. The apparent lack of technological complexity in traditional elements does not imply that they are less efficient than novel materials. For such purpose we shall contrast the features and design involved in the determination of light redirecting elements such as lightshelves, with those composed of modified materials.

A window without adequate treatment can be a source of potential comfort problems, due to the excessive contrast between the zone close to the window and that in the opposite end of the office room. Uncontrolled penetration of solar radiation can increase thermal loads during summer, producing an additional load to air-conditioning systems.

2.5.1 Lighting Redirection Systems and Their Calculation

Lighting redirection systems are used in situations where it is desirable to extend the reach of natural lighting, balancing contrast, elevating illuminance levels at the back of a room, and in the process save energy by reduced usage of artificial lighting when this is zoned between different areas of the room (Ochoa and Capeluto 2006). Two widely used systems include lightshelves and anidolic systems (Fig. 2.15).

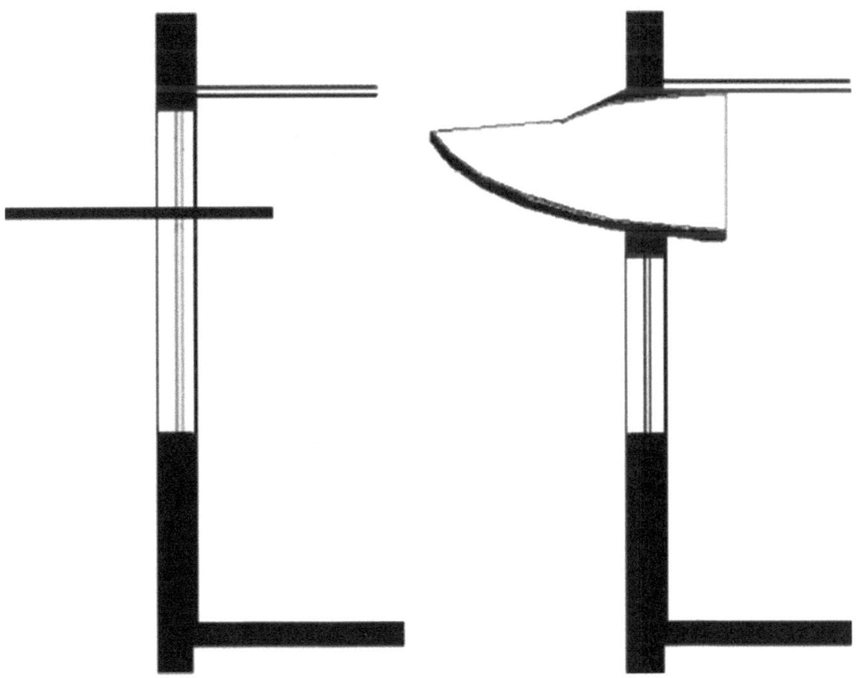

Fig. 2.15 Sections for typical configuration of a lightshelf (*left*) and an anidolic concentrator (*right*)

In the first one, a reflective element is placed horizontally in the clerestory section of the window. Usually this element has a very shiny top surface for light bouncing. On the second element, two parabolic mirrors are placed substituting the clerestory part. They have an entrance and exit glazing, and their mirrors are placed in such way that they can bounce light rays to the desired back portion of the room ceiling, where it reflects back to the inner space (LESO-PB 2008).

For a system with fixed geometry, determining its most appropriate geometry can be a fundamental decision, which can also become limiting in its future performance. Nevertheless, certain elements can be taken into account. Determining the geometry of the daylighting systems can be done by using the angle of incidence of solar rays for an intermediate position for shading during summer and winter time.

The evaluation of the systems can be done through specialized programs for daylight simulation such as Radiance (LBNL 2016), although the subject will be dealt in more depth in Chap. 4. A complete study will consider adapting evaluations for light quality and quantity, since studying one variable, such as illuminance, can be misleading if human factors are not incorporated. It is a known fact that users will react to correct the amount of light entering a space in order to achieve visual comfort, and can even sabotage such systems. Therefore, glare analysis should also be incorporated.

The expected results from the analysis could be similar to Fig. 2.16, which shows the results for glare simulations. The results provide an assessment of the

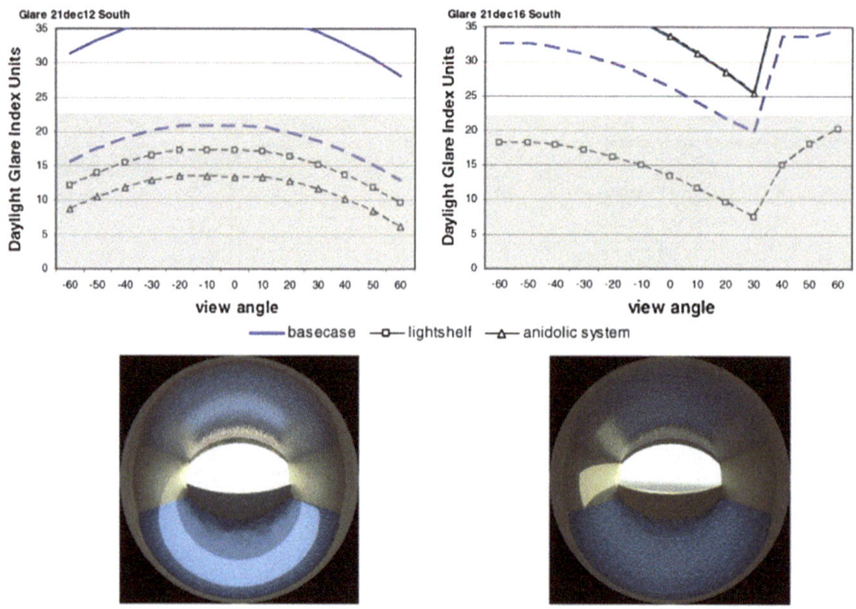

Fig. 2.16 Glare analysis results in analytical (*top*) and graphical form (*below*). *Gray* area in the graph represents the comfort zone. *Dashed lines* indicate the use of horizontal blinds

contribution of the different solutions to reduce glare probability, compared to a basecase without solar protection. The user would then, based on these results, evaluate the most favorable model according to the particular situation. Therefore, although the elements do not have any moving parts, careful and intelligent study is still needed prior to their application in the intelligent façade, as they have significant impact on its morphology.

Summarizing this example, the generic methodology that is followed for these type of devices is the following:

(a) Definition of project objectives.
(b) Selection of potential candidate elements to fulfill the objectives.
(c) Dimensioning and testing (through computer simulation or prototyping).
(d) Feedback from testing results and final detailing.
(e) On-site implementation.
(f) Monitoring and adjustment.

On-site implementation for these devices refers to the actual constructive phase. For bespoke designs, there might need to be an intermediate stage for full-sized prototyping. This stage is different from the one made for energy performance, and is used for verification of structural and manufacturing features.

2.5.2 Smart Shading Systems and Their Implementation Process

The development of materials with enhanced properties through different chemical, physical or fabrication processes, has become one of the most significant innovations for the construction industry (Lazarovich 2012). Although there are precedents in the architectural use of innovative materials, such as in photovoltaic cells or photochromic glazing, new possibilities are opened by the blurring of previously known material limits.

Nevertheless the excitement they can produce for new possibilities, the horizons that can be explored, and the amount of research invested in them, it is hard to find an adequate definition for smart materials. A satisfactory statement describes them as "materials whose properties can be varied controllably in response to changes in their environment" (Bhavsar and Vaidya 2008).

This makes the study of their properties to be closely related to developments in other fields, such as chemistry and materials engineering. Nevertheless, currently the transmission of information is not as fluid as it should be between engineers and architects. Knowledge generated in one field needs a translation for usability in the other. Some reasons behind this gap are the different aims followed by each discipline. Architects require definite properties to conceptualize a space; material engineers usually are researching how to incorporate new features to their products, even though they might not have clear which ones.

To exemplify this process, we summarize the process for a smart façade shading system (Lazarovich 2012). Its development began by defining the design priorities. These consisted of providing a modular stackable system, which incorporated geometrical movement for optimal blocking of solar radiation throughout the year, while providing a visual focus element.

These priorities were then set forth as purposes the system had to comply with, namely providing blockage of solar radiation in different stages, operation under different temperature ranges in a geographical location, geometry change, maximum foldability for storage when not required, ease of maintenance, and usage of familiar technological analogies to carry out the response mechanisms. These requirements were expressed in an objective data format.

During the exploratory phase, different possibilities were examined. The solutions were collected as a catalogue of materials that detailed their main physical properties. Information was then presented in the catalogue in a search system that is useful for architects (Lazarovich et al. 2012). Examination of this catalogue of solutions revealed that for the desired system, a combination of nickel-titanium shape memory alloy metal springs, which acts as sensor and thermal actuator, and canvas shading materials could fit the requirements. It also provides for use of smart materials with low installation cost and requirements.

The next step required a detailed exploration of different options for the shape memory alloy and the folding systems. For the smart metal section, it required filtering the many options available for shape memory alloys, and knowledge on how to specify the required components. For the folding canvas section, a detailed study was made on several alternatives of folds and mechanisms that would allow meeting the requirements. Exploration was made on conceptual mockups with shape polymers that did not have the final materials, due to its high cost.

When a suitable solution was determined, it was then tested experimentally on a proof of concept. The prototype used a scaled down sample of the shape memory alloy and 3D printed components for mechanism and the canvas support. The proposed shading component's opening/closing mechanism is based on equilibrium between two compression springs, as mentioned previously: the shape memory spring and a simple steel spring. When ones compression force is greater than the other, the component changes its state (Fig. 2.17).

This setup was tested indoors for its response at different temperature ranges. The final pattern mimicked flower petals opening and closing when heat was applied and when it cooled down, respectively. The pattern is dynamic since each hexagonal component on the building envelope can sense its particular exterior temperature or radiation level and respond directly (Fig. 2.18).

An initial evaluation of the shading potential and the radiation exposure on a glass façade was done, evaluating the incident radiation on the façade during summer. The results show that, in the opened state of the components, there is no direct exposure to the sun. In the semi-opened state of the components, the façade absorbs half of the radiation that is absorbed in the closed state, in which most of the façade is exposed to the sun's radiation. In order to allow the use of diffuse daylight inside the office spaces, materials like fabrics or polymers with different

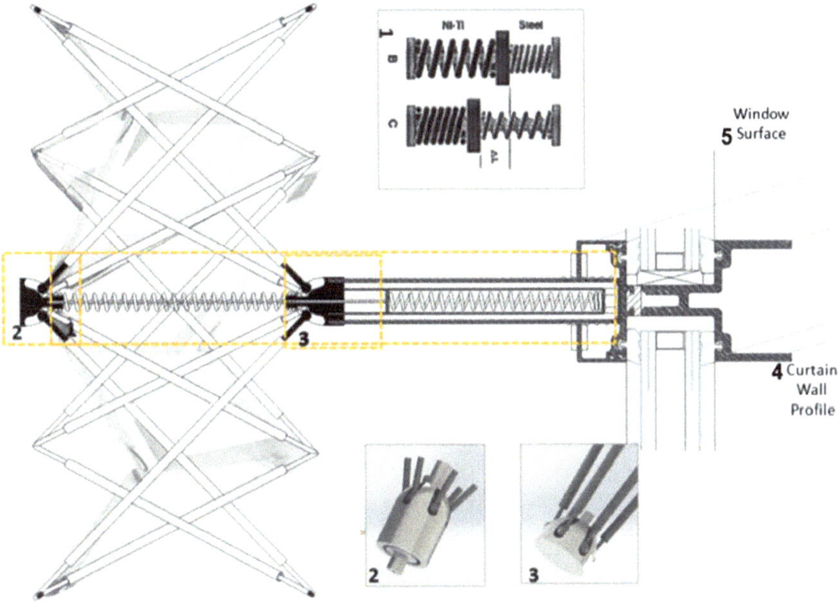

Fig. 2.17 Sections and views for the smart shade using memory alloys. *Image credit* © Courtesy Noy Lazarovich

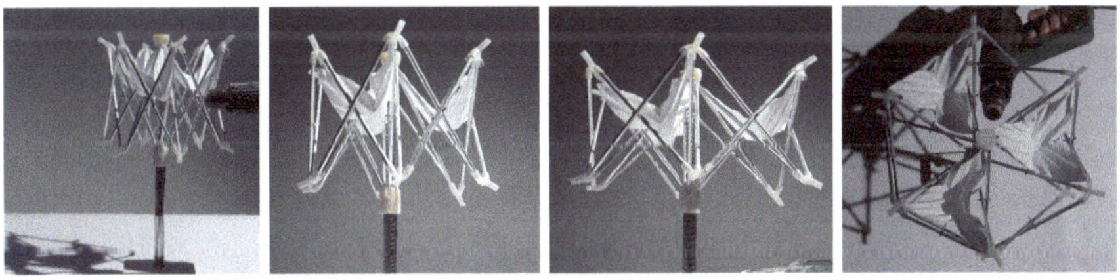

Fig. 2.18 Prototype testing of a smart shade using memory alloys. Screenshots from video. *Image credit* © Courtesy Noy Lazarovich

levels of opacity can be used according to the specific needs in each section of the building envelope. As a conclusion from this example, it can be said that the development stages for smart materials can be generalized for most of them. They would include the following steps:

(a) Definition of general project objectives.
(b) Specification of requirements from the material(s).
(c) Search for existing solutions, based on the requirements.
(d) Selection of candidate materials for meeting the requirements.
(e) Testing of prototype systems that will use the material(s).

(f) Calculation of expected impacts, based on general objectives and results from prototype testing.
(g) Full-scale implementation.
(h) Monitoring and adjustment.

However, carrying out each stage is practically a custom process, since each case is different and will have its own requirements. There is also the risk that requirements asked from a smart material have not been developed yet. Or in the worst case scenario, that the desired material does not exist.

The last step of the suggested process is also composed of a series of sub-stages, since many smart materials are at this stage expensive for architectural use and only available for small components. It is expected that a study on their industrialization process might be required, in order to meet the desired characteristics for a complete building project.

Therefore, caution must be taken when using these types of smart material actuators, as depending on them for carrying out all the desired envelope functions will make the building dependent on a given technological development.

References

Anderson D (2010) Humanizing the hospital: design lessons from a Finnish sanatorium. CMAJ 182:E535–E537. doi:10.1503/cmaj.090075

Arcspace (2007) United States federal building morphosis. Available at http://www.arcspace.com/features/morphosis/united-states-federal-building. Accessed June 2016

Arqa Website (2011) Edificio Consorcio Santiago: 14 años despues. Available at: http://arqa.com/arquitectura/edificio-consorcio-santiago-14-anos-despues.html. Accessed May 2016

Arup (2016) SolarLeaf-Bioreactor façade. Available at http://www.arup.com/Projects/SolarLeaf.aspx. Accessed Jan 2016

Bhavsar R, Vaidya NY (eds) (2008) Intelligence in novel materials. Oilfield review, Spring 2008, pp 32–41

Bozsaky D (2010) The historical development of thermal insulation materials. Periodica Polytechnica: Architecture 41:49–56. doi:10.3311/pp.ar.2010-2.02

Canadian Encyclopaedia (2016) Architectural history: early first nations. http://www.thecanadianencyclopedia.ca/en/article/architectural-history-early-first-nations/. Accessed Jan 2016

Compagno A (2002) Intelligent glass facades: materials, practice, design, 5th edn. Birkhauser, Basel

CTBUH—Council on Tall Buildings and Urban Habitat (2011) KfW Weterkade Frankfurt. Available at http://www.ctbuh.org/TallBuildings/FeaturedTallBuildings/kfWWestarkadeFrankfurt/tabid/2896/language/en-US/Default.aspx. Accessed Jan 2016

Curtis WJR (1994) Le Corbusier: ideas and forms. Phaidon, London

Diffenderfer R (2005) Electronic devices: systems & applications. Thomson Delmar Learning, Clifton Park

Fortmeyer R, Linn C (2013) Kinetic architecture: design for active envelopes. Images Publishing, Mulgrave

Giurgiutiu V (2007) Concepts of adaptronic structures. In: Adaptronics and smart structures: basics, materials, design, and applications. Springer, Berlin

Hanser DA (2006) Architecture of France. Greenwood Press, Westport

Henson S (2012) Computer game industry shaping user interface design for intelligent buildings, finds IMS Research. FM Link, 5 Dec 2012. Available at: http://fmlink.com/articles/computer-game-industry-shaping-user-interface-design-for-intelligent-buildings-finds-ims-research/. Accessed May 2016

Herzog T (ed) (2010) Oskar von Miller Forum. Hirmer Verlag, Munich

Jelle BP, Gustavsen A, Baetens R (2010) The path to the high performance thermal building insulation materials and solutions of tomorrow. J Build Phys34:99–123

Keoke ED, Porterfield MK (2003) American Indian contributions to the world. Checkmark Books, New York

KNX (2016) KNX protocol introduction. Available at www.knx.org. Accessed Jan 2016

Lazarovich N (2012) Smart material systems for high performance building envelopes: guide-lines for smart material implementation in the design of climatically responsive building envelopes. Dissertation, Technion Israel Institute of Technology

Lazarovich N, Capeluto IG, Silverstein MS (2012) SMARTerials for high performance buildings. In: Bartolo et al (eds) VRAP 2011—advanced research in virtual and rapid prototyping, 5th international conference, Leiria, Portugal. Taylor and Francis, London

LBNL-Lawrence Berkeley National Laboratory (2006) Advancement of electrochromic windows. Document CEC-500-2006-052. Available at: http://www2.lbl.gov/Science-Articles/Archive/sabl/2007/Jan/Advance-EC-Windows.pdf. Accessed Jan 2016

LBNL-Lawrence Berkeley National Laboratory (2016) Radiance synthetic imaging system. Available at http://radsite.lbl.gov/radiance/. Accessed May 2016

LBNL-Lawrence Berkeley National Laboratory (2016a) Daylighting the new york times headquarters building. Available at: https://facades.lbl.gov/newyorktimes/newyorktimes.htm. (Accessed October 2016)

LESO-PB (2008) Anidolic systems at solar energy and building physics laboratory, EPFL. Available at http://madd.epfl.ch/e/research_dl_anidolic_results.html. Accessed May 2016

National Board of Antiquities (2005) Nomination of Paimio Hospital for inclusion in the world heritage list. Available at: http://www.nba.fi/fi/File/410/nomination-of-paimio-hospital.pdf. Accessed Jan 2016

New York Times (1983) The 'intelligent' buildings. New York Times digital archive. Available at http://www.nytimes.com/1983/12/01/business/the-intelligent-buildings.html. Accessed Jan 2016

Ochoa CE, Capeluto IG (2006) Evaluating visual comfort and performance of three natural lighting systems for deep office buildings in highly luminous climates. Build Environ 41:1128–1135

Reinhart CF, Voss K (2003) Monitoring manual control of electric light and blinds. National Research Council Canada, Publication 45701

Senott RS (ed) (2004) Encyclopedia of 20th century architecture, vol 1. Fitzroy Dearborn, London

Technical Preservation Services (2016) Preservation of historical adobe buildings. https://www.nps.gov/TPS/how-to-preserve/briefs/5-adobe-buildings.htm. Accessed May 2016

Towler J (2011) World market for intelligent environmental building control and contrasts with North America. LonMark training session, 31 Jan 2011. Available at https://www.lonmark.org/connection/presentations/2011/Q1/AHR/04+Intelligent+Building+Controls–BSRIA.pdf. Accessed May 2016

WICONA (2016) Autarc façade. Available at http://www.wicona.com/en/de/archive/Range-of-systems/Facades/Autarc_facade/. Accessed Jan 2016

Wigginton M, Harris J (2002) Intelligent skins. Architectural Press, Oxford

Chapter 3
Design Considerations

3.1 Stages of the Architectural Design Process of an Intelligent Envelope and Impact on Performance

In order to understand better how to apply intelligence to buildings, it is necessary to take a look at how architects and building planners work. It is of interest to see how they develop the idea of a façade and assign a given technology to it.

The architectural design of envelopes is one of the aspects taken into account during the overall design process of a building. Although many architects work under the "black box" approach to the design activity, several steps of the process are recognizable. They are applied by most designers independent of the outcome.

The architectural design process moves iteratively. Although there is no unified theory describing the steps, it is generally agreed that it begins from abstract ideas and concepts, progressing towards more specific definitions. Different interpretations are given by design theoreticians on how architects pass from one step to the other (Brawne 2003).

The earliest design stages deal with general definition of broad elements such as location of building volumes within the plot, volume massing and orientation, allocation of spaces within them, and in some cases, the inclusion of vague ideas about the façade and roof, and perhaps some ideas about the technological means to be used to achieve the design objectives. At present, many of those ideas are drawn out in the sketch form that distinguishes the architectural labor from that of other construction professionals. These early drawing representations of ideas for the building shape and envelope are elaborated further as more information is collected about the project, and as additional feedback is given on the feasibility of carrying out those ideas. More specifics are given to each detail on each iteration, and elements such as the façade and roof take shape into a more recognizable form.

Second-stage iterations can see the definition of characteristic elements such as establishment of structural elements, decisions on window modulation and decisions on roof type and shape. If the project is complex enough, different

G. Capeluto and C.E. Ochoa, *Intelligent Envelopes for High-Performance Buildings*, Green Energy and Technology, DOI 10.1007/978-3-319-39255-4_3

practitioners might have been assigned to carry out the final design of separate aspects of the building. Depending on the procedures of each architectural office, communication might have been established with façade or environmental consultants about the expected goals of the project and technologies to be applied.

The next stage sees the development of definitive ideas that are assigned realistic materials and dimensions. The data is detailed enough to be shared with consultants that might be proposing specific technological solutions based on that information. Final (or close to final, as some practices can attest) decisions on technology and design are then brought together to the specification and document-production phase.

3.2 Architectural Design Strategies for Intelligent Envelopes

We have described how the overall design process and the paths taken by the design team influence the performance of intelligent envelopes. Any design team that considers envelope intelligence at the start of the design process will be faced with decisions about the types of strategies to follow. At the same time, they might have to balance other influential factors unrelated to climatic strategies. Examples include total construction costs, construction schedule and even marketing. An extreme case occurs when a given orientation that is "negative" for energy consumption might define how well the project is sold (for example, if this coincides with panoramic views, etc.).

The main choices available for the application of different technologies are passive, active and hybrid strategies. A description is given here of their main characteristics:

Passive design strategies

These strategies denote the exclusive use of bioclimatic principles that achieve energy performance goals in the simplest way, through the use of good design methods such as providing adequate orientations, defining satisfactory building shapes and depths, performing studies on openings' size, designing volumes according to solar and daylight rights, promoting natural ventilation, etc. The elements used respond to climatic requirements over long periods of time and depend on weather characteristics to be activated. Examples include façade or slab elements with heavy thermal mass, basic vacuum-glazed windows, inlets for natural and cross ventilation, buoyant flows in ventilation chimneys, fixed shading elements and lightshelves (Fig. 3.1), etc.

The intelligence provided to the envelope using these strategies resides principally in the early design stages. However, it needs good specification if there are many passive strategies combined in a way that border with an active system. On one hand, users might not intervene on the operation of certain systems, but on the

Fig. 3.1 Fixed vertical shading elements in the San Francisco Federal Building by Morphosis. *Photo credit* I. Guedi Capeluto

other, they might need to undergo some training in order to understand specialized manual operations of devices such as two-sided blinds, for example.

Many of the elements used in passive strategies have been used and tested in good traditional construction, such as shutters, ventilation grilles, solar chimneys, etc. Their application is recognized but will not be discussed in detail as there is abundant literature on the subject. Although they might seem basic and well-known, research continues on them in order to improve their design, combination, application and capability (Vijayaraghavan 2016).

Active design strategies

An exclusively active design strategy would refer to the sole application of elements with high degrees of technological sophistication, usually through moving parts or enhancement of their physical properties. Examples include blinds capable of independent rotation of each blade, enhanced insulation systems (Kosny et al. 2008), electrochromic glazing, etc.

Intelligence in this type of strategy depends exclusively on element performance. This might lead designers to trust their confidence on the good functioning of envelope elements for project success. This dependence provides arguments to opponents of highly "active" intelligent buildings, pointing out that they can be risky during times of failure or maintenance. The degree of complexity present in each of the elements can leave little room for planning a backup strategy.

Active design strategies can be considered both at the start and end of the design process. However, if taken at the end of the design process without complementing basic climatic design considerations, the effects of intelligence will be significantly reduced, as demonstrated in Sect. 3.3. Therefore, it is imperative to consider them from the start of the process.

Examples of these strategies can be found sometimes in buildings that want to showcase the benefits of a given technology that begins to be introduced in the construction market. The value of these buildings is for benchmarking, yet dependence on one technology strategy is not advisable.

Hybrid design strategies

As the name implies, these strategies include a combination of both passive and active ones. It can be said that this is most of what is found in practice for intelligent envelopes, representing integral planning of a building and careful component selection. Design decisions are the result of a process in which climate considerations have been taken from the beginning. Both morphological and element selections are chosen to work together towards a common goal.

Successful implementation depends on supplementing each technology so that passive and active strategies enhance each other. Intelligence is provided to the building envelope in the planning stage and through selection of intelligent features.

This can only be achieved if envelope intelligence is considered from the beginning of the design process, with technical considerations being given proper importance during morphology design and component detailing. The combination of different appropriate strategies is a positive approach to cases in which budget might be tight, but it is still desired by the building owner to provide the project with visible active features.

Furthermore, good planning of a hybrid design strategy will help avoid total failure of building systems, in case one of the active elements cannot be used due to need for maintenance or breakdown. In such case, performance will not be as effective as when both strategies are working together, but alternate courses of action can be available to building owners in case they want to upgrade or replace.

Final users might still need training or explanations on how the building works, but conscious design should also include simple and intuitive operation as part of the details.

Intelligent envelopes planned in this way offer flexibility and convenience that exclusively passive buildings cannot provide. An example is the ability to adapt to temporary changes in short time spans, custom operation of given elements, ability to provide personalized comfort, etc.

3.3 Impact of Design Decisions Due to the Addition of Intelligence

This account of the design process is closely related with intelligent envelopes, since the performance outcome has been already largely determined in the very first design stages (Hari 2001). For instance, deciding on a sub-optimal massing orientation, can compromise the thermal and visual conditions inside the building and require additional treatment. It has been demonstrated that the yield from applying intelligent elements chosen at the final stages is not as high as if passive design and climatic considerations have been applied from the earlier design stages (Ochoa and Capeluto 2008).

These outcomes were found after energy modelling a case study for an open-plan office building located in the Mediterranean city of Haifa, Israel. The application of intelligent facades in hot climates has been barely studied, while there is a growing body of knowledge for those in cold areas. Replicating solutions developed for cold climates creates high energy consumption levels and troublesome working conditions. For the case study, an evaluation was made of three variations series. In all of them, the basecase is an office module that faces west, with a window area of 20 % of the floor area, and has no shading devices (pure, flat form). For simplicity, in this basecase artificial lights are on during all working hours. The orientation is unfavorable but has prime real estate value in the city, since it provides direct views to the sea and faces coastal breezes. It represents a common case where high importance was given to factors such as views.

The first series of variations, numbered 1–8 (Table 3.1) improves the basecase by gradual addition of *active features*. The original design intentions are preserved (pure, clean and flat form), while the performance of individual elements is observed. In this series, various highly active elements are used with no change in the general volume: reflective blinds for radiation and glare control operating automatically at a fixed angle, stepped lighting controls, and forced night ventilation driven by electric fan with specific external temperature setpoints, as well as changes of glazing from double glazing clear to low emissivity glazing.

Table 3.1 Studied cases for assessing impact of design decisions due to incorporation of intelligence

Series	Case	Orientation	Glazing	Blind type/activation	Light control	Night Vent.	Lightshelf	Code in graphs
Active features only	1 (basecase)	West	DGC	None	on	–	–	w-Cl-Ib-on
	2	West	DGC	Internal/radiation	on	–	–	wW-Cl-IbR-on
	3	West	DGC	Internal/glare + radiation	on	–	–	w-Cl-IbGR-on
	4	West	DGC	External/glare + radiation	on	–	–	w-Cl-EbGR-on
	5	West	DGC	Internal/glare + radiation	on	F&S	–	w-Cl-IbGR-on-aV
	6	West	DGC	Internal/glare + radiation	st	F&S	–	w-Cl-IbGR-st-aV
	7	West	LE	Internal/glare + radiation	st	F&S	–	w-Le-IbGR-st-aV
	8	West	LE	External/glare + radiation	st	F&S	–	w-Le-EbGR-st-aV
Passive design only (low energy)	9	West	DGC	External/fixed	st	–	–	w-Cl-Eb-st
	10	West	DGC	Internal/fixed	st	–	–	w-Cl-Ib-st
	11	West	DGC	Internal/fixed	st	nV	–	w-Cl-Ib-st-nV
	12	West	Le	Internal/fixed	st	nV	–	w-Le-Ib-st-nV
	13	West	Le	External/fixed	st	nV	–	w-Le-Eb-st-nV
	14	South	Le	None	st	nV	–	s-Le-==-st-nV
	15	South	Le	Internal/fixed	st	nV	–	s-Le-Ib-st-nV
	16	South	Le	Internal/fixed	st	nV	1.0 m single	s-Le-Ib-st-nV-Ls

(continued)

Table 3.1 (continued)

Series	Case	Orientation	Glazing	Blind type/activation	Light control	Night Vent.	Lightshelf	Code in graphs
Passive design and active features	17	South	Le	Internal/radiation	st	nV	1.0 m single	s-Le-IbR-st-nV-Ls
	18	South	Le	Internal/glare + radiation	st	nV	1.0 m single	s-Le-IbGR-st-nV-Ls
	19 (10 m deep)	South	Le	Internal/glare + radiation	st	nV	1.0 m single	s-Le-IbGR-st-nV-Ls [10 m]
	20	South	Le	Internal/glare	D	aV	0.5 m upper 0.5 m lower	s-Le-IbG-Di-aV-HsLs
	21	West	Le	Internal/glare	D	aV	0.5 m upper 0.5 m lower	w-Le-IbG-Di-aV-HsLs
	22	North	Le	None	D	aV	0.5 m upper 1.0 m lower	n-Le-==-Di-aV-HsLl
	23	North	Le	None	D	aV	–	n-Le-==-Di-aV-===

Notes
DGC Double glazing clear (3 mm + 6 mm + 3 mm) VT = 0.81 SHGC = 0.76
Le Low emissivity (6 mm + 6 mm + 6 mm) VT = 0.68 SHGC = 0.42
on Always on
Di Dimmer
st Stepped
aV Fan and setpoints
nV Natural ventilation

The second series, numbered 9–16, adds improvements to the basecase using *intelligent passive design* strategies. It features low-energy elements implementing basic climatic principles for the particular location. The elements in this series are: fixed shading, stepped lighting controls passive natural night ventilation, and low-emissivity glazing. In the second series a fundamental change is also tested by orienting the facade to the South, adding a lightshelf to an "optimized" passive module.

The third series in the table, numbered 17–23, shows the combination of *intelligent passive design strategies with active features*. Variations 20, 21 and 22 are optimized alternatives for different orientations. They are derived from prescriptive requirements of the Israeli energy rating of buildings standard (IIS 2011) to obtain the highest energy-efficiency grades that it can award to an office project.

Results from the study are shown in Figs. 3.2, 3.3 and 3.4. From them it can be seen that designers can expect to save 8–40 % if they assign intelligent elements at the end of the design process, compared to 55 % and more if they choose to design with appropriate climatic design strategies from the beginning. Therefore, intelligence for the envelope, for a higher performance output, needs to be considered during all design stages.

Nevertheless, providing adequate performance guidance during the early design stages when design freedom has a high degree of change, is the only time when changes can be interactive and made easily (Capeluto 2013). Later on, changes become more expensive if not impossible to make. Another factor that complicates matters and needs to be considered, is that performance of different alternatives can only be accurately evaluated when alternatives have been clearly defined and measurements put in place. This is because of the current nature of most evaluation tools used by consultants, which require accurate numbers at a stage when these are not available (Ochoa and Capeluto 2009).

Design experience can also be a help or a hindrance towards achieving high-performance envelopes. It has been observed that although design experience correlates with design performance, experienced designers seem to do less well when solving atypical problems (Roan 2008). Also, certain "design cultures" within some professional practices can be influenced by the still prevailing misconception that an intelligent technology by itself will be able to solve everything; thus those design cultures would separate design from technology application.

Designing intelligent envelopes, at the moment, can still be considered an atypical design problem, for which many options and variables with mutual influence are available. The novelty of the field makes it easier to experiment with new technologies that might still lack market establishment or whose results have not been verified in large scale use for each specific climate.

One of the outcomes of the design activity and its influence on performance is that architects need to have at least basic knowledge of climatic design principles, or they need to be assessed through adequate tools that in theory should be their companions for the early design stages. More on that subject will be dealt with in Chap. 4, Design Tools.

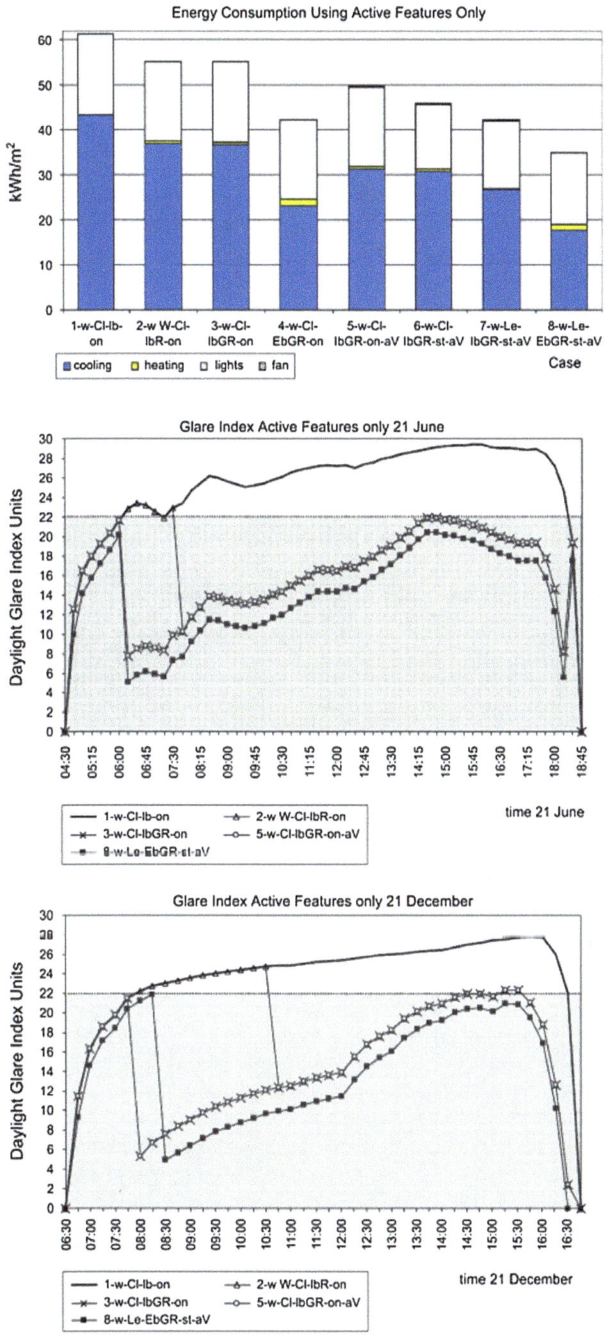

Fig. 3.2 Energy and visual comfort analysis for design decisions using active features only for the city of Haifa, Israel

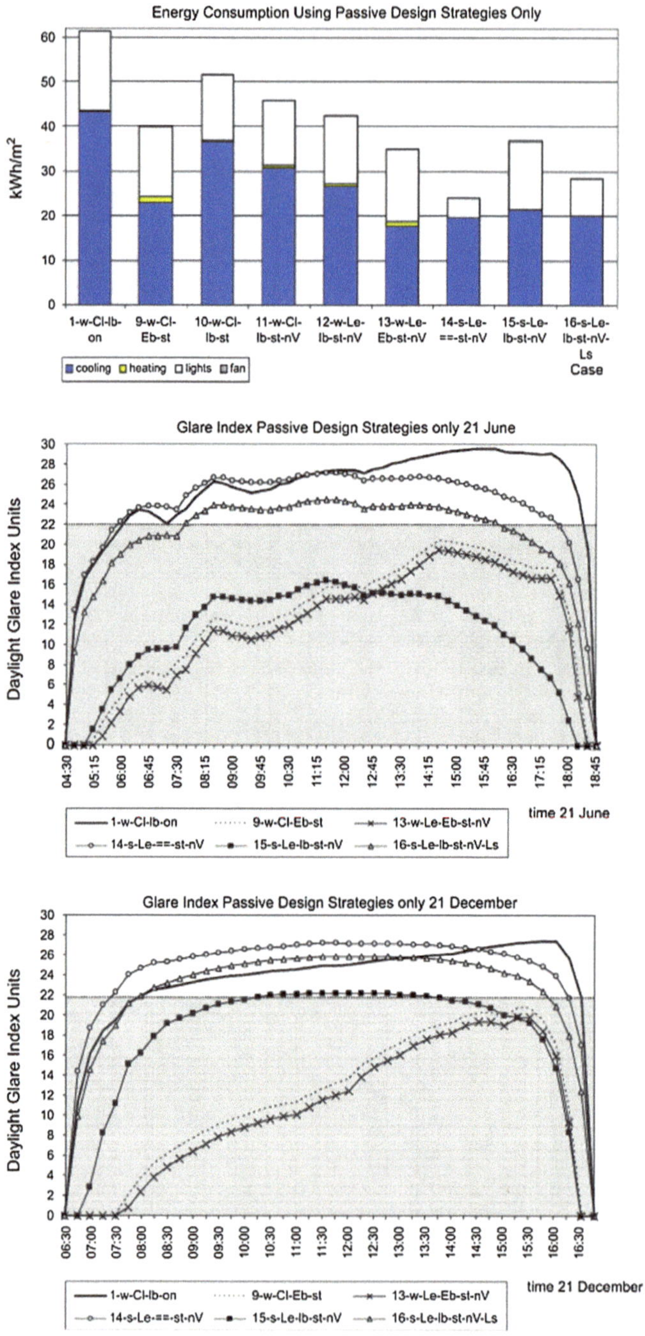

Fig. 3.3 Energy and visual comfort analysis for design decisions using passive features only for the city of Haifa, Israel

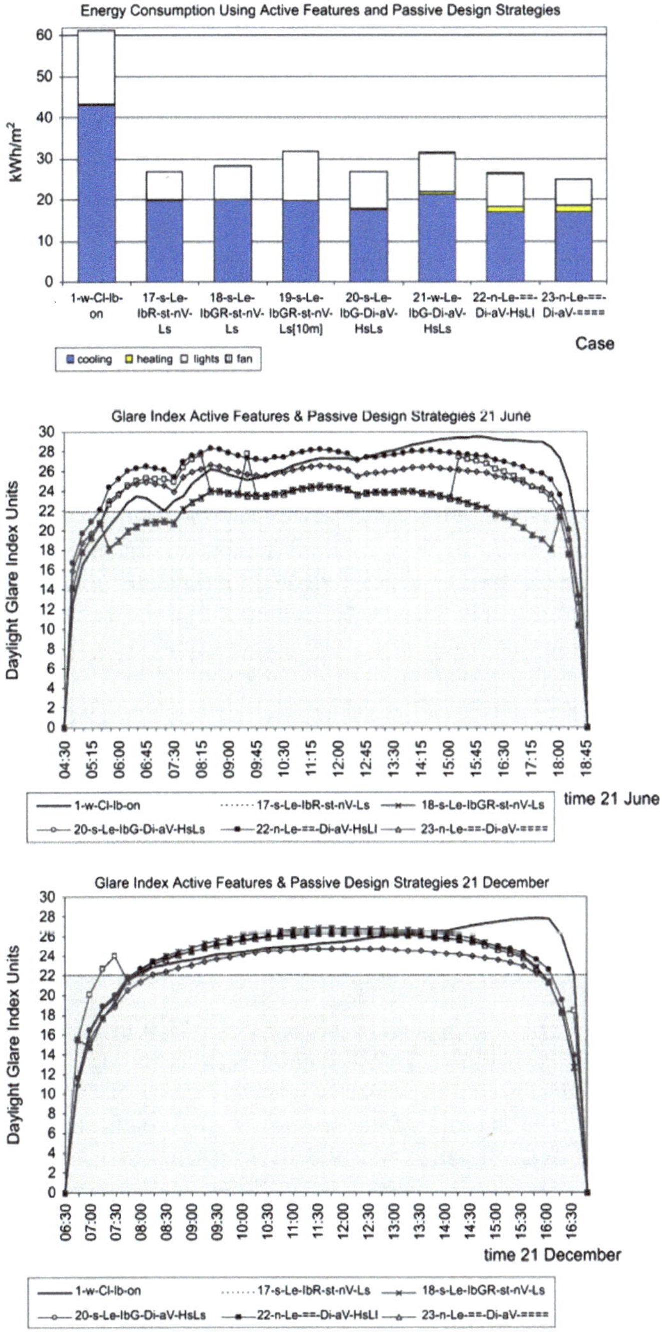

Fig. 3.4 Energy and visual comfort analysis for design decisions using passive features and active elements for the city of Haifa, Israel

3.4 Difference Between the Architectural Design Process of Intelligent Envelopes for New and Retrofit Buildings

Increasingly, architects are being faced with the task of upgrading current buildings, besides designing new ones. This is being done not only because of economic considerations by building owners, but also as part of wider international efforts to reduce energy consumption and greenhouse gas emissions. Since a very large percentage of the building stock worldwide has been erected before the introduction of high-performance energy standards, extensive energy retrofits are needed to bring it to a low-energy consumption level (Atanasiu et al. 2011).

A feasible way to achieve this goal is by retrofitting the building envelope, as it is relatively easier to replace or add constituting elements such as insulation or windows.

It is important to note that in retrofits, envelope upgrading takes a prominent place in the design process since it might be among the few key elements that can be replaced from scratch or can be fit with appropriate intelligent technologies. A retrofit envelope can help to improve lighting, insulation, ventilation and infiltration (Dall'O et al. 2012) in order to reach comfort levels and significantly reduce energy consumption and emissions. And not less important, envelope retrofit prominently signals that a building has been renewed.

The design process of a smart envelope that has as objective energy retrofit is slightly different from that of a new building. While in the new design all variables are open to experimentation and change (including its building shape and orientation), the upgrading of a building starts from a set of constraints where some of them cannot be modified. For example, façade orientations are already given. In some cases, opaque areas on facades and roofs cannot be modified, this would add additional constraints on the amount of technologies that can be placed on each area.

However, the architectural design approach to the retrofit process still uses similar methods to that of a new building. Priorities of what needs to be achieved are set out clearly from the beginning through dialogue with the clients or statements found in the design brief. To avoid unforeseen technical difficulties that the existing structure could offer, there is also relatively closer collaboration with the consultancy team from the start of the process. Another particularity of a retrofit design process is that the design as well as delivery schedules can suffer unexpected changes if new but decisive information is found throughout project execution. An example of the latter can occur if during construction it is discovered that the structural strength of original supporting elements does not reach the calculated levels (Parrish and Regnier 2013; Schlueter and Thesseling 2008).

Two main options have been identified for the execution of envelope energy retrofits (Konstantinou and Knaack 2011): the first is total element replacement, where the building is stripped of its original walls and roof and new components are built from the beginning. In this case, designers have more freedom to choose new intelligent elements but construction times are longer and more disruptive to inhabitants. It also needs careful structural integration between the new façade and the existing slabs.

The second option is that of modular additions, and has been used mostly when a single passive strategy is pursued. For example, when the analysis determines that upgrading insulation or replacing windows brings significant energy savings compared to the original condition. Although it can be structurally independent of the building, it can add significant thickness to existing outer walls and roofs. The advantage is that construction times are reduced and there is less demolition work, making it a preferred method for residential retrofitting where it is desirable to reduce the impact on current inhabitants.

Research efforts have been also directed to produce modular panels for envelope energy retrofits that can contain more than a single strategy, to provide maximum flexibility and adapt for different climates. They can be provided with smart materials (Hughes 2013) or some degree of artificial intelligence for active elements (Bresaer 2016).

3.5 How to Form Design Strategies

In this section we will deal with the relevant aspects towards the formation of hybrid design strategies. Since each situation is different, general guidelines are provided for the design team to consider. The overall objective that should be strived is to reach occupant comfort through minimal energy expenditure. However, each case requires particular study before implementing a given set of solutions. For this reason, we present here a brief revision of climate strategies that are generalized for two main climate families: hot and cold.

Replicating solutions developed for a specific climate zone, which are totally different from the project one, may create unacceptable energy consumption levels and troublesome occupancy conditions. Independently from where designs are conceived, to be appropriate for the local climate, they should be based on proven building physics principles, which are reviewed in this section, together with the initial inventory of appropriate intelligent technologies presented in Chap. 2.

In the following sub-sections, we exemplify the main points that need to be taken into account for adequate development of design strategies that will be applied in the design of intelligent envelopes:

(A) **Climate principles**

Considering climate design principles are useful to keep in mind user comfort and energy use reduction. They are specific for each location and define the interactions between building and outer environment.

Strategies can be derived from examination of the biophysical effects of the environment inside a building (Givoni 1998). They show how changes in the climate affect temperature levels, thermal comfort, air velocity, relative humidity and solar radiation absorption. With these factors in mind, climate strategies can be grouped in the following groups:

1. Heat management, collection and storage
2. Ventilation for comfort and air quality
3. Daylight (and sunlight) admission and control.

In order to understand better these strategies, a graphical contrast is made between hot and cold climate requirements (Fig. 3.5). Actual locations might have a mixture of both characteristics during certain times of the year, therefore it is important to perform a good climate analysis beforehand.

In cold climates, as an example, heat collection and storage is essential. Air intake is also limited for the same reasons, but enough ventilation must be provided to control indoor air quality for occupants' health. Due to short daylight times during winter and potential cloudy conditions in other seasons, maximum daylight penetration is desired mainly in public buildings (Hutcheon and Handegord 1983).

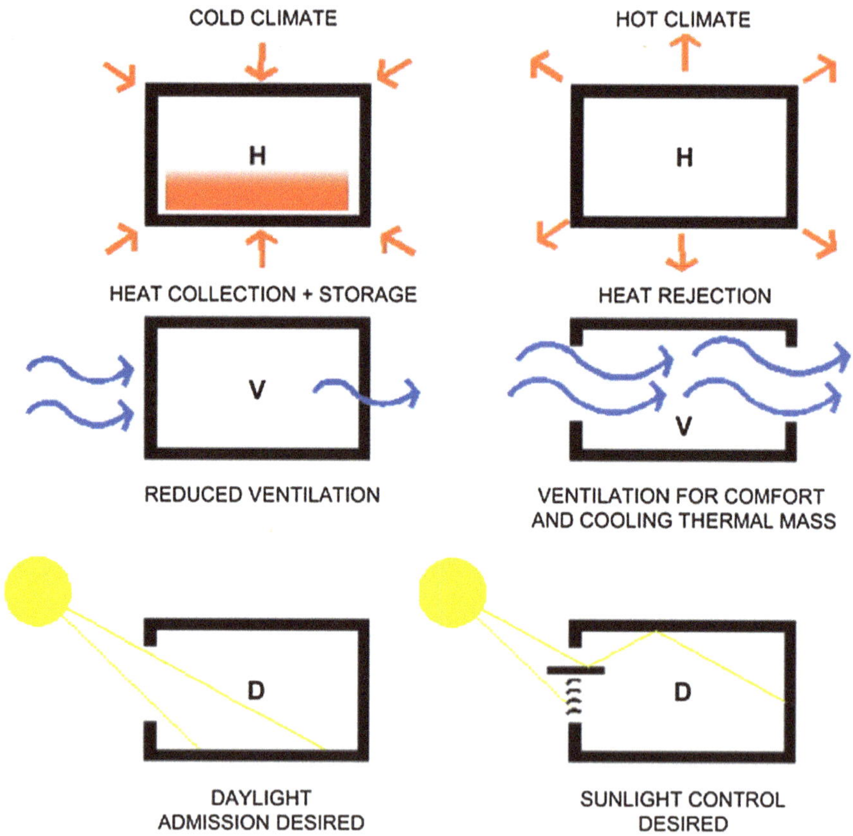

Fig. 3.5 Contrast between the main climatic strategies for cold and hot climates

Fig. 3.6 Double façade for heating and ventilation. Oscar von Miller Forum. *Photo credit* Carlos E. Ochoa

An implementation of these ideas is shown in Fig. 3.6, where a double façade serves to the objectives of heat collection, ventilation as well as daylight regulation.

On the other hand, in hot climates heat must be excluded as much as possible during warm seasons to prevent high internal temperatures. Relative humidity must be controlled internally, in order to provide human comfort; this can be achieved through adequate ventilation means.

Daylight and sunlight penetration has to be carefully managed in order to avoid increasing internal temperatures and to reach comfortable internal visual levels. Figure 3.7 shows how this strategy is implemented in the CH2 building of Melbourne, Australia. It depicts a motorized shading device that turns the wood shading elements as required.

Although climatic building design strategies are important, it cannot be stated that any one of them has a higher priority than the rest. An existing situation might determine which one has to be pursued more actively, helping to achieve better energy performance. But the large number of elements existing in the market today, and the short time to choose from them makes it necessary for designers to use particular methodologies and specialized tools that help them identify the best combinations for different situations.

Fig. 3.7 Application of sunlight control strategy through a motorized shading device in the CH2 building, Melbourne, Australia. *Photo credit* I. Guedi Capeluto

(B) Importance of site analysis

Identifying climate principles is useful to carry out site analysis that can provide precise data for testing different options for strategy combinations. The subject of specific tools will be dealt in more detail in Chap. 4.

One of the first steps, which might seem obvious, is to consider traditional site analysis for topography, orientations, wind patterns, access and views. This will help establish the potential placement of energy-generating technologies such as building integrated photovoltaics (BIPV) and where to place opaque and transparent façade elements and their protective features.

An important derivation of solar analysis is not only the characteristic sun path diagrams, but also the influence on shaping building volume. The study of solar rights will be reflected in the possibility to allow solar passive heating during winter and avoidance of shaded areas on other buildings or adjacent open areas due to the project itself. Clever solar right planning ensures both access to solar radiation and high built area densities (Capeluto et al. 2006).

Solar analysis also includes observation of neighboring obstructions such as surrounding taller buildings, significant vegetation (high trees that are deciduous or not), obstructions created by topography such as hills or mountains that cast shadows on the site during certain times of the day and compromise the potential of passive solar design and daylight use. This also implies a study that considers year-long shadow behavior, which is readily done through different kinds of tools for both on-site study and off-site prediction.

Comfort studies based on climate conditions are based on different theoretical models that examine weather characteristics and correlate them with experimental observations on acceptance and performance in different combinations of temperature and relative humidity (RH). These combinations can be reflected in a bio-climatic or psychrometric chart that reflects the type and duration that conceptual strategies have to be followed during the year in order to restore human comfort (Olgyay1963).

Implications from the recommendations for site analysis include the need to visit the actual site to collect information, and to obtain reliable weather data that can be used as input for site analysis tools.

(C) Economic considerations

A financial analysis focusing on preliminary construction expenditures usually determines the final inclusion or not of certain technological elements. Typically, main lines of initial cost will be divided between traditional passive technologies, active technologies and experimental technologies if these are used. In this way becomes important to identify climate strategies and how they will be solved, as it simplifies the task of observing those basic indispensable courses of action, and possible substitutes for active technologies if they are found to have high initial or maintenance costs that cannot be covered by the building owners.

This is a traditional method to carry out basic economic analysis, which can determine carrying out the building activity or not. Nevertheless, there are other factors to be taken into account.

One such factor is the payback period, in which benefits given by the technologies will reach initial investment levels and in some cases even start generating revenues in terms of savings. However, calculations need to take into account the technology yields and rates for maintenance and failure which can be hard to predict in the long term in particular for experimental technologies (Kasprowicz and Schulz 2015). Financial periods for return-of-investment vary, and might also depend on the time frame that building stakeholders determine to be acceptable.

In some countries, subsidies and credits are available as a way to encourage the application of energy-saving technologies that have been proved or for those that have a high yield but also high initial cost. Financial construction analysts have to make sure that their project qualifies under the terms of such aids.

The economic impact of quite novel technologies is hard to assess in particular if there have not been many examples, but this also becomes a circle in which cautious investors might not want to apply innovative elements due to the uncertainty of yields over time and return of investment. This makes field tests important in order to understand how these new technologies can be applied in practical conditions and their potential savings (Kosny et al. 2013).

(D) **Technological limitations and user acceptance**

Designers have to be realistic, when it comes to the limitations posed by intelligent features in building envelopes. As mentioned in Chap. 1, intelligent building envelopes are rational agents. They will only respond to what they are programmed to do. Such limitations are placed beforehand by the responsible team for design strategies through the instructions that reflect climate strategies.

Therefore, responsibility falls to designers in terms of providing sufficient courses of action for these elements, within the limits of reason and common usage. Active technologies in particular might need regular maintenance to ensure their long-term functioning. Most of them are now equipped with sensors and software that may provide timely maintenance announcements. During these down periods suitable strategies can be placed for continuation of user comfort. An example can be increasing ventilation if blinds are retracted and sun-light is entering during maintenance times, or placing them in a fixed shading mode.

It must be remembered that moving systems tend to wear down with time and need to be repaired or replaced. The feasibility of a long-term maintenance program should be taken into consideration when selecting elements and proposing budgets. Even if such programs are in place, systems that are too complicated or unusual might eventually be abandoned after many years due to lack of spare parts or replacement by new versions of that system.

Operational and behavioral aspects have to be considered together in order to provide an acceptable product to end users, facility managers and owners. In this sense, user interaction can be studied in the design phase through different operational scenarios. The approach can help identify different intelligent system interactions, and the role of each active and passive element.

At the user level, many of the scenarios studied by the design team can be offered as part of the series of presets to be found in the operation interface. These interfaces should be understandable even by the most basic level of user, and should suggest common usages and alternatives of control. Another important feature of ideal interfaces is that they should be able to cater for individual preferences as much as possible, such as desired temperature, lighting levels, and ventilation volume and air speed.

A system that reacts too fast to weather changes can also be a detriment towards user acceptance. This effect has been observed particularly in smart blinds that

Fig. 3.8 Consorcio Vida building in Santiago, Chile, as an example of integration with its urban context and acceptance by users and the city. *Photo credit* I. Guedi Capeluto

operate almost instantaneously to changing light levels, resulting in confused users. Nowadays it is usual to provide blinds a certain deployment lag which can be useful when clouds are passing, for example.

An aspect that requires additional study by designers considering envelope intelligence is the integration of the project with its surroundings. This can be considered a test for acceptance within the city or urban area. Since the envelope provides an image that is part of the city, element arrangement should also strive to contribute to its context with a positive, durable image.

An example is given in Fig. 3.8, where the vines of the Consorcio Vida building break the prevalent monotony of nearby office blocks adding green surfaces to the city. The use of living vegetable materials in the vertical plane also becomes a city landmark, recognizable by inhabitants and visitors alike. Changing colors throughout the year also provide a fluctuating panorama that brings variety to the experience of users and passers-by.

(E) **Additional considerations for envelope retrofit**

For existing buildings that will undergo envelope retrofit with intelligent elements, it is necessary to do a detailed assessment, in order to decide on the appropriate course of action. The evaluation will take into account a variety of aspects that have to be considered to increase the feasibility of improvements. They include the percentage of energy improvement, increase in occupant comfort, new technology interactions with the existing building, structural viability, and economic costs through a payback analysis. Many times, an additional aspect that is overlooked is the execution of studies on the potential influence or disturbance on neighboring buildings, or surrounding micro-climate. This can occur, for instance, when additional floors are added to a building that will undergo renovations.

Just as a new building is seen as a whole, thought and planning has to be given to areas beyond envelope renovations. Refurbishment of internal spaces and conditioning systems should be thought to contribute to energy and comfort performance. In the case of office buildings, new surface finishes have to be specified for improved light distribution, and for residential ones there should also be a program for replacement of acclimatization systems for energy efficient ones

3.6 The Self-shading Building Envelope

In this section we present the design process for a self-shading envelope (Capeluto 2003), as an example of the form principles involved when designing an intelligent envelope. It is used to highlight many of the factors involved in determining its shape and enhancing its performance. They can be applied to other types of facades and roofs as well.

During the early design stages for a building, the architect deals with general geometrical factors related to its shape. These factors include building plan proportions, height in relation to street dimensions, and facade orientation. In these early stages, solar potential of the building and from its surrounding areas are determined. This process will assure appropriate shading during critical hours of summer days, and exposure of facades and surrounding sidewalks to the winter sun.

There are several examples where designers tried to shape buildings with the objective of enhancing their thermal performance. The building form was shaped in a manner that allows insolation in winter to achieve passive heating of the building, and partial self-protection from solar radiation during a warmer period. Examples of such buildings are the Bank of Israel in Jerusalem (A. and E. Sharon Architects), the municipality building of Bat Yam (Hecker, Sharon and Neuman), the Blue Cross and Blue Shield of Connecticut Building in North Haven, Connecticut (Ellenzweig Associates Inc.), and the city halls of Boston (Kallmann, McKinnel and Knowles) and Tempe (Michael and Kemper Goodwin).

In these examples, an upside-down pyramidal shaped structure was created. This helps to achieve self-shaded facades during a certain period in summer. Buildings with an inverted pyramidal geometry may increase their roof area, such as in the Tempe city hall.

The application of such model is relevant for buildings where design decisions require a self-shading geometry for a given period. The self-shaded period is the time during which the building envelope has to avoid the penetration of direct solar radiation into the building.

The designer must determine the length of this self-shading period, according to climatic and programmatic considerations. Since the method presented here is limited to self-shading solutions, the designer can also determine the required period by considering combining a partial self-shading solution with additional dynamic shading devices.

Although these issues can be checked manually, using sun diagrams or protractors, it is a rather long and complex task that has to be done for all the building facades and for different times of the year. A more efficient method is the use of the solar collection envelope. The term *solar envelopes* define the space of solutions for the determination of a design that considers either insolation or solar shading. These envelopes define possible solutions to a given problem.

There are two types of solar envelopes (Capeluto and Shaviv 2001): The first one is the *solar collection envelope* (SCE), which presents the lowest possible location of windows and passive solar collectors on the elevations of a building, allowing their exposure to the sun during a given period at winter, thus avoiding shading by existing buildings (Fig. 3.9). The second type, the *solar rights envelope* (SRE) defines of the volume inside which one can build, without violating the solar rights of neighboring buildings and avoiding being shaded by them. For the self-shading envelopes described in this section, determination of the solar collection envelope was studied.

The solar collection envelope is defined for each project according to a definition of the self-shading period required, which is specific for each site. The critical date and time that shape SCE geometry may be different for each elevation depending on its orientation.

Calculations are carried out for the 21st day of every month, and for every hour during the predetermined required period. For each hour during this period, the azimuth and altitude of the sun are calculated.

Conceptually, the examined area has to be divided into a fine orthogonal mesh placed at the desired building height area, and the maximum height value from the zenith to the vertices of each of the mesh components, to the ground surface, need to be calculated. Joining all the point elements in space will determine the SCE surface.

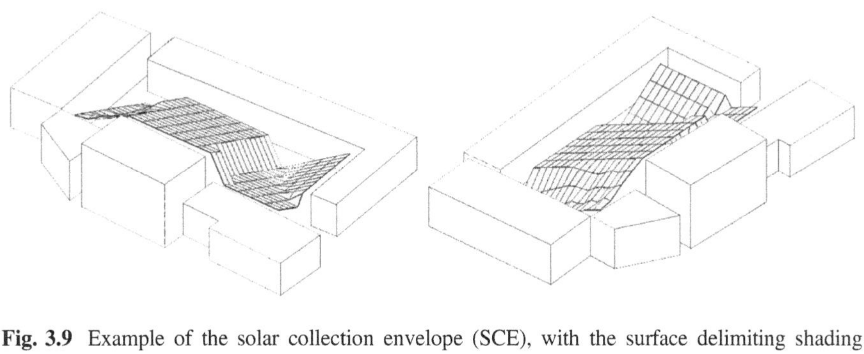

Fig. 3.9 Example of the solar collection envelope (SCE), with the surface delimiting shading influence of existing buildings during winter months

This process of determining the protection angle and the building shape is not a simple task, particularly for non-rectangular building plans, or when the North axis is not parallel to the external walls. The required shading period for a window in a given orientation should be calculated as a result of parametric analysis, using dynamic hourly energy simulation models. Such analysis for different shading requirements gives the energy consumption for heating, cooling and lighting. After finishing calculation of the required shading period, it is possible to generate a building geometry that fulfills these requirements.

The longer the required self-shading period, the larger will be the wall inclination, in particular for east and west orientations. This also means a lower resulting building volume with a disproportionate roof area exposed to the sun. In order to decrease wall inclination and therefore roof area, designers may consider additional means such as deeper windows and the use of additional dynamic shading devices on the affected facades.

As an example, we present a study done for the city of Jerusalem (32°N 35°E), which was carried out using the SustArc simulation model (Capeluto and Shaviv idem). The calculations were carried out for a shading period based on the summer months (May to September) between 10.00 and 14.00 solar time. As the solar path is symmetrical for 22 June, the resulting shading period will be wider, including March and April. The daily self-shading period was limited to four hours around noon, in order to avoid very inclined facades.

As a way to prevent significant solar gains before 10:00 and after 14:00 in summer, additional shading devices should be considered by the designer. In cases when the aim is to protect the building during all the working hours, it may be required to run the model using the local standard time.

From Fig. 3.10, it can be observed that for the studied case, the SCE obtained is symmetrical with respect to the North-South axis. This explains that the resulting East and West facades result with the same inclination of 34 degrees from the zenith. However, this is not the case for the South and North facades.

On the southern elevation, the required inclination is 31 degrees from the zenith, which is significant. For the North elevation there is no need to incline the wall. The reason for this is that, as expected for this location, the sun does not impact the north facade during the required period.

However, it is important to emphasize that in certain cases it may be necessary to protect the North elevation, depending on the location and required period for solar protection (in terms of months and hours). In that case, an additional inclination for the North facade will be suggested for the resulting solar collection envelope. In this case, designers might also consider implementing a tilt angle to the North wall, as a manner to continue the same morphology, or they can consider adding traditional methods for shading using other devices.

Figure 3.10 also shows the southern and eastern facades of the solar collection envelope. SustArc calculated the envelope for all the possible heights, measured from the roof downwards, resulting in a closed surface. The designer can determine the shape and profile of each façade according to the required number of floors. A visual examination is presented at the bottom of Fig. 3.9 for evaluation of the

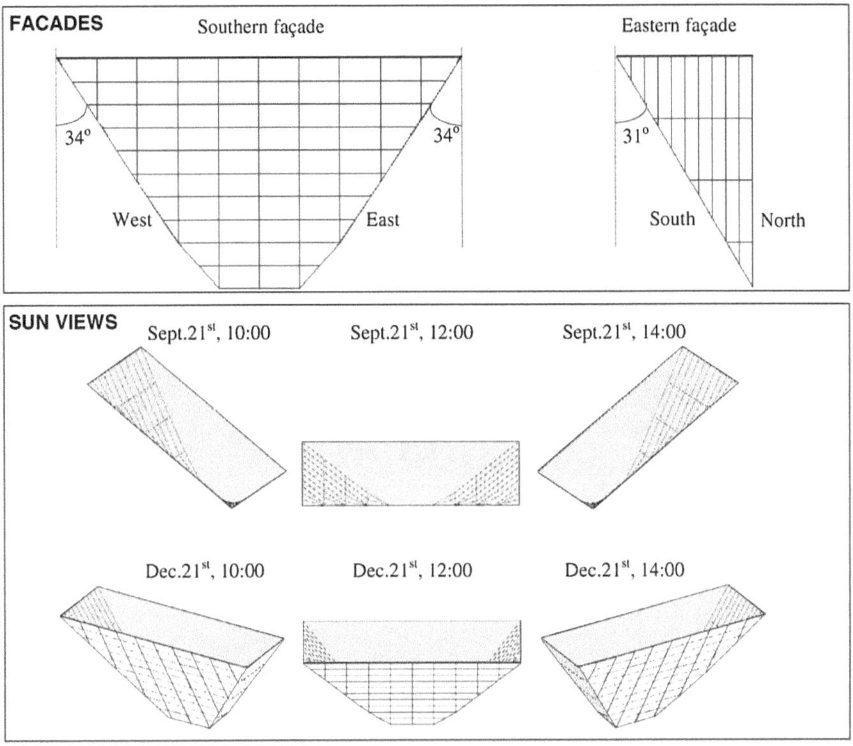

Fig. 3.10 Example of a SCE for a North-South oriented building in Jerusalem generated using the SustArc design tool

obtained envelope. The volumes are presented in isometric view as seen from the solar position at a given time. As seen in the figure, the presented views were done for two specific dates:

The first date was 21 September, as a limit for the required shaded period, when the sun is in the lowest position for this period. The second date was 21 December, which represents a winter day. This second date was calculated to verify possible solar exposure of the building during the winter period.

According to this envelope, a second stage was performed in the study: two different alternatives for the stepping pattern were proposed in the design of the southern facade of the studied building, as shown in Fig. 3.11. As mentioned before, the required façade inclination for this orientation and required shading period is 31°. The two proposed alternatives have windows mounted in a vertical plane. An alternative that wants to follow more directly the form would consider installing the windows in the plane of the sloped surface, which is also possible. In this case, the designer must be aware that such decision will increase considerably the total window area.

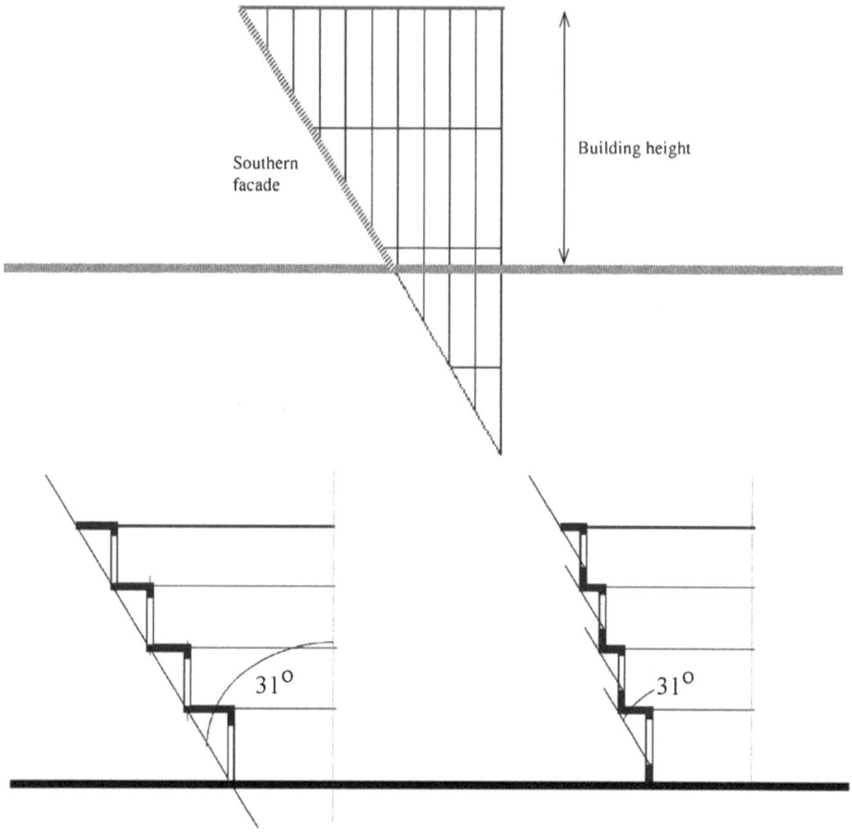

Fig. 3.11 Two pattern alternatives for a South self-shading façade located in Jerusalem following the SCE

To confirm that this type of volume is useful for energy performance, solar radiation and energy simulations were performed. The simulation model used calculates the amount of short wave radiation that hits each window for every hour, according to the wall inclination and orientation, as seen in Fig. 3.12. This figure shows the incident direct and diffuse solar radiation during the summer months in the studied location.

Results from the radiation simulation show the following results:

(a) In the East and West facing facade, the window plane sloped at 34° eliminates the penetration of direct solar radiation from 10:00 to 14:00 h, which is the period for which the self-shading envelope was generated. However, in the early morning hours, for the east orientation, and late afternoon hours for the west orientation, there is still some direct radiation reaching the building facade. Nevertheless, it is important to remark that even in those cases, direct radiation hitting the sloped facade is significantly lower than in a completely vertical facade.

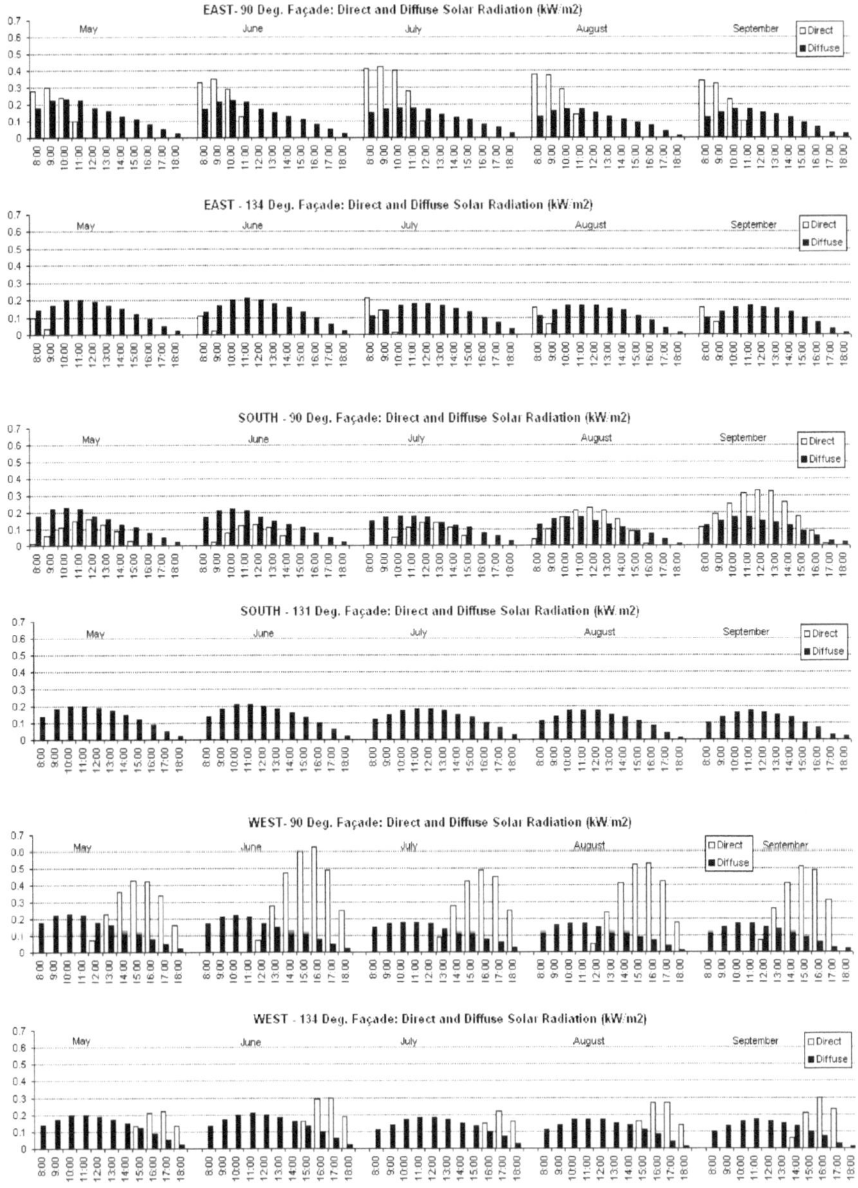

Fig. 3.12 Direct and diffuse solar radiation for south, east and west orientation according to facade inclination, during summer months in Jerusalem

(b) In the South facing facade, the window plane sloped at 31° avoids penetration of direct solar radiation during all working hours. Moreover, the reduction of diffuse radiation is very small, allowing the use of daylight and promoting visual comfort.

Energy consumption was also calculated. Figure 3.13 shows the comparison for cooling in the analyzed design alternatives. The following results can be mentioned:

(1) For all window orientations, there is an improvement in energy performance of the building that was designed according to the self-shading envelope. However, this improvement is more significant in East and West orientations, than for offices located in the South.
(2) The use of internal blinds is known as helpful for vertical facades facing to the South. However, for South-oriented windows using the self-shading envelope, show that there is no need to provide additional internal blinds for thermal improvement purposes.
(3) In East and West orientations, adding internal blinds improves the energy performance of the building space. This is due the contribution of the blinds during the hours when direct radiation penetrates into the office through the window. Moreover, in these cases a better distribution of daylight and avoidance of glare is to be achieved as a result of using internal blinds.
(4) The enlargement of the roof due to the implementation of facade inclination does not produce a deterioration of the overall energy performance of the building.
(5) Using high-performance low-emissivity glazing, with a shading coefficient of 0.44, reduces energy consumption for cooling with a similar effect to self-shading. However, the glazing type selection should be done considering the whole building energy use including electric lighting.

The design potential of these solar collection envelopes makes them a useful element that designers can consider in the early stages of the design process, which is useful to achieve an energy-saving design of the building. Accurate determination of the self-shading envelope can be obtained in a more practical way using a model

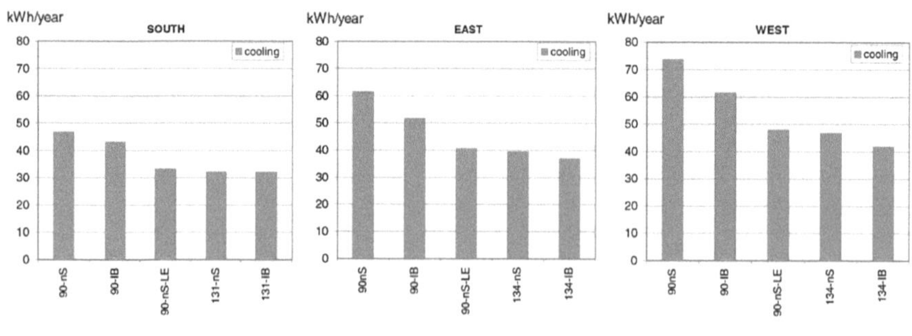

Fig. 3.13 Influence of facade inclination, glazing type, and shading on energy consumption for an upper floor office, Jerusalem

such as SustArc. Through its use, designers and planners may get immediate graphic and numeric output for the design of different geometric variations based on the SCE, in an accurate and easy way.

Computer tools such as the one mentioned, can also be used at an urban level. The objective of using the tool at this scale has as purpose determining street profiles, obtaining profiles for shadowed sidewalks and facades during a certain period at summer, but exposing them to the sun during winter. The use of this design tool does not lead to a unique final design, hence the freedom of the architect is not limited. An extensive discussion on the use of design tools on the design process will be presented in Chap. 4.

Even though, the method presented here has a number of limitations. In the first place, it is applicable only to buildings where the designer wants to obtain a self-shading envelope. Secondly, the required shading period must be limited, particularly during the early morning and late afternoon hours, to avoid extremely sloped walls. According to this method, limits need to be set by the designer, which can lead to certain errors in estimation. Since solar gains during these hours could be significant in summer, an additional shading program must be determined for each of the facades.

3.7 Considering Intelligence for a Good Design

The previous sections have shown that in order to maximize the energy potential of intelligent envelopes, these must be considered early in the architectural design process, from concept and form to details. Such process must have a multi-disciplinary feedback through energy, technology and economic analysis.

As an integrative design process, it involves a degree of complexity that is different from a "regular" building envelope. Part of the intricacy lies in the high number of elements involved, and the interaction of both active and passive features. If experimental elements are involved, there is also the degree of uncertainty in the performance of novel elements. Such factor can be solved through prototype experimentation.

The design process, in order to produce intelligent envelopes, must also be viewed through a methodological scope to include the steps of perception, reasoning and action. This is a departure from the traditional "black box" architectural design processes, where sometimes form for the sake of form is its end and objective. It also differs from achieving functional performance as the only goal (making the envelope devoid of any creative spirit), since for integral reasoning all aspects need to be considered and provided with adequate solutions for each case.

An integral design process for intelligent envelopes also involves bringing down preconceived notions about the envelopes themselves. They are not a catalog of kinetic elements (as they are usually viewed by some), but apply in dynamic manner the principles required for good climatic design. It is also important to note that they are not the sole responsibility of consultants brought at the final stage of the process, but must be considered from the start. The exploration of environmental factors as a form generator has been retaken as an organized activity the last decades, and there are still many areas left for investigation.

A successful intelligent building, as seen from the considerations presented in this chapter, is not a patchwork of smart active features that are expected to work seamlessly only from their properties. It needs to be the product of a conscious design process that incorporates intelligence in all its stages, including the schematic early ones, while taking advantage of technological innovations and the interactions between each other.

Climatic building strategies are important on their own, but it cannot be considered that a particular one has an absolute priority over others. Each particular site will determine which one has to be pursued more vigorously in order to achieve minimal energy consumption and maximum comfort through element selection.

The presented case study, for the methodology used to design a self-shading envelope, demonstrates the early stages of the design process involved in detailing its characteristics. Such process involves delimiting certain variables, through the use of specialized tools for evaluation and design of the element. In the case presented previously, they focused on building form for energy performance. Further development and detail of the envelope regarding materials, systems etc. should be expected as the design process advances.

It is also recognized that the complexity of intelligent envelopes can be staggering at the beginning of the design process, when there is not enough information available and plenty of ways to proceed are open. Therefore, designers need tools for early stage evaluation of the condition. These tools are presented in Chap. 4.

References

Atanasiu B, Despret C, Economidou M, Maio J, Nolte I, Rapf O (2011) Europe's buildings under the microscope: a country-by-country review of the energy performance of buildings. Buildings Performance Institute Europe

Brawne M (2003) Architectural thought: the design process and the expectant eye. Architectural Press, Oxford

Bresaer Project (2016) Bresaer project website www.bresaer.eu. Accessed Feb 2016

Capeluto IG (2003) Energy performance of the self-shading building envelope. Energy Build 35:327–336

Capeluto IG (2013) Design tools for asking questions: evaluation vs. generation. EAAE Trans Archit Educ 61:602–613

Capeluto IG, Shaviv E (2001) On the use of the solar volume for determining the urban fabric. Sol Energy 70:275–280

Capeluto IG, Yezioro A, Bleiberg T, Shaviv E (2006) Solar rights in the design of urban spaces. In: PLEA2006, the 23rd conference on passive and low energy architecture, Geneva, Switzerland, 6–8 Sept

Dall'O G, Galante A, Pasetti G (2012) A methodology for evaluating the potential energy savings of retrofitting residential building stocks. Sustain Cities Soc 4:12–21

Givoni B (1998) Climate considerations in building and urban design. Van Nostrand Reinhold, New York

Hari A (2001) Integrated, customer-driven, conceptual design methodology. Dissertation, Technion—Israel Institute of Technology

Hughes C (ed) (2013) EeB PPP project review: FP7-funded projects under the 2010, 2011 and 2012 calls, November 2013. Electronic document. https://ec.europa.eu/research/industrial_technologies/pdf/eeb-ppp-project-review-2010-2011-2012_en.pdf. Accessed Feb 2016

Hutcheon NB, Handegord GOP (1983) Building science for a cold climate. Wiley, Toronto

IIS—Israel Institute of Standards (2011) SI 5282, Part 2-Energy rating of buildings: office buildings (in Hebrew)

Kasprowicz R, Schulz C (2015) Availability-based payback method for energy efficiency measures. Proc CIRP 29:710–715

Konstantinou T, Knaack U (2011) Refurbishment of residential buildings: a design approach to energy-efficiency upgrades. Proc Eng 21:666–675

Kosny J, Yarbrough DW, Miller WA (2008) Use of PCM enhanced insulation in the building envelope. In: BEST 1 Proceedings, Building for energy efficiency and durability at the crossroads, 10–12 June, Minneapolis

Kosny J, Shukla N, Fallahi A (2013) Cost analysis of simple phase change material-enhanced building envelopes in Southern U.S. climates. Building technologies program, U.S. Department of Energy. Report DOE/GO-102013-3692

Ochoa CE, Capeluto IG (2008) Strategic decision-making for intelligent buildings: comparative impact of passive design strategies and active features in a hot climate. Build Environ 43: 1829–1839

Ochoa CE, Capeluto IG (2009) Advice tool for early design stages of intelligent facades based on energy and visual comfort approach. Energy Build 41:480–488

Olgyay V (1963) Design with climate: bioclimatic approach to architectural regionalism, 4th edn. Princeton University Press, Princeton, New Jersey

Parrish K, Regnier C (2013) Proposed design process for deep energy savings in commercial building retrofit projects. J Archit Eng 19:71–80

Roan ET (2008) Early-stage engineering design: the designer, the object of design, and design context. Dissertation, Massachusetts Institute of Technology

Schlueter A, Thesseling F (2008) Balancing design and performance in building retrofitting: a case study based on parametric modeling. In: Proceedings ACADIA 08 Silicon + Skin > Biological processes and computation, Minneapolis, 16–19 Oct 2008

Vijayaraghavan K (2016) Green roofs: a critical review on the role of components, benefits, limitations and trends. Renew Sustain Energy Rev 57:740–752

Chapter 4
Design Tools

4.1 Brief Overview of Existing Design Tools

As seen in previous chapters, there is a multitude of variables that need evaluation and assessment when designing intelligent envelopes, but for which their management requires specialized calculation methods. Sometimes it might also be required to do some testing previous to full scale implementation in order to verify data.

Designers can resort to a series of methods that will assist them in the evaluation of different design options for intelligent envelopes. These methods have been abstracted as tools, which are available in different formats such as manual tools, computer software or printed guidelines. However, at the moment there is no "universal tool" that can be used as a design companion from start to finish in the design process of the intelligent envelope. Instead, a series of design aids are used, appraising a variety of characteristics or specialize in the analysis of a single variable. Therefore, designers and consultants need to be knowledgeable in a number of approaches and procedures according to their needs. These will help them in order to apply their judgment to the results provided by tools that evaluate different aspects used for planning intelligence in a building.

It must also be noted that a large percentage of existing design tools and aids are suitable only for the advanced stages of the design process (Ochoa and Capeluto 2009). Many design tools present important disadvantages for efficient use during the early design stages: most are not user friendly, many require exact numbers in order to provide usable answers, and above all, previous experience is needed on the subject in order to interpret the expected results.

This makes them more usable for "regular" building design than for the incorporation of extremely advanced features. However, they are required as part of the "arsenal" of any design team attempting to incorporate intelligence in the design of the building envelope.

G. Capeluto and C.E. Ochoa, *Intelligent Envelopes for High-Performance Buildings*,
Green Energy and Technology, DOI 10.1007/978-3-319-39255-4_4

In-depth analysis of each building aspect is complex enough to require faster methodologies than those offered by traditional hand calculations and formulas. The action of producing intelligent buildings is currently intricate enough, needing the use of specialized tools. Their use should allow both a smooth constructive process and fluid information-sharing, from the planning to the execution phase. This favors the availability of computer-based tools as the most widely media to achieve these objectives. Currently they are widely available through different dissemination channels (plug-ins, custom-made tools, paid, free-of-charge, etc.).

4.2 Tool Classification

The classification offered in this chapter does not aim to be an exhaustive list of all tools that have been created for the design and analysis of intelligent envelopes, but instead provides representative cases that illustrate their main characteristics. Tools will be divided into the following categories according to their main usage within the design and consultancy process.

- Provision of basic guidelines
- Generation tools
- Single-aspect evaluation tools
- Whole-building simulation
- Other tools (including hybrid ones)
- Design suggestion tools.

The number of tools that have been developed (both in the past and present) is extremely large, and new ones are being released each day. Therefore, the chapter will not focus on mentioning names or trying to be an exhaustive literature review of everything that is available, but concentrate remarking their main features, advantages and limitations.

Instead, the classification and examples mentioned in the chapter will provide readers with a brief description of the main characteristics for each type. This has the objective to form criteria upon which users can employ these tools in the development of intelligent envelopes. All these studied groups can be considered useful for design and energy analysis of envelopes with varying degrees of intelligence, with their utilization for such objectives being highlighted.

4.3 Tools Providing Basic Guidelines

Many climatic design principles have been abstracted as guides to facilitate their implementation by designers. Initially they were featured in print as nomograms, rulers, graphs, etc. Currently, new media channels allow for a wider distribution, and they can be found as software packages or dedicated webpages. Their main characteristic is the almost linear provision of answers to a single aspect in the design such as comfort level, shading area, etc. through mostly graphic means. A shared characteristic among many of these tools, is that their instructions are relatively simple to follow, enabling teaching of building physics and climatic design to architects and engineers.

They can be used as the basis to formulate draft general strategies that have to be followed by intelligent envelopes. Another important use they have is that they enable studying the main features in the project climate, helping identify those façade or roof elements that can satisfy specific comfort requirements or be translated to energy saving strategies. At the beginning of the design process they prove useful to perform site studies, as they can help to show those times when neighboring obstructions shade a given façade or areas of a roof surface.

Examples include a variety of implementations using the psychrometric chart for bioclimatic design (Olgyay 1963; Givoni 1998). The chart makes it easier to determine climatic strategies to be followed, by comparing weather characteristics from a given location such as temperature and humidity, with paired limits of the same items for human comfort and performance. This comparison is fulfilled based on several comfort models through execution of a certain number of actions, indicated in the more complete versions of the chart.

The early steps of the design of intelligent envelope elements can benefit by using these indications, as the goals can be programmed into them according to each season of the year. Originally found only in print version, numerous software implementations can be obtained for different platforms. They provide interactive guidance on the type of strategies and some even show practical examples. Some of them include desktop-based such as Climate Consultant (Milne 2016), adaptation for use in popular programs such as Excel (Kelsey 2015) and even mobile platform applications (ASHRAE 2016).

Other basic guidelines available include solar path diagrams and nomograms (Szokolay 2007). Their importance for intelligent envelopes resides in determining solar position at a given time or period, which can affect the design of shading devices and lightshelves, window sizing, placement and installation of photovoltaic panels (Solar Pathfinder 2016), solar trackers (Prinsloo and Dobson 2014), solar heaters, etc. Guidelines for passive solar building design, with implementation examples, can also be found (Yezioro and Shaviv 1996; Yezioro 2009).

There is also a large amount of media in which they are presented, such as custom printing from the web (UO SRML 2016), as tutorials (CLEAR 2016), for direct online use (Agafonkin 2016), we can also find them for complex testing for different environmental variables inside other evaluation tools, such as Ladybug (Roudsari 2016). This tool provides a graphical-based way to examine results and use them in a 3D modelling environment.

Another important issue to be considered during design of the building envelope is posed by solar volumetric considerations which ensures that the entire project or critical parts of it will have solar access or block the solar access of others. A series of tools exist to determine solar and daylight rights according to design aims and

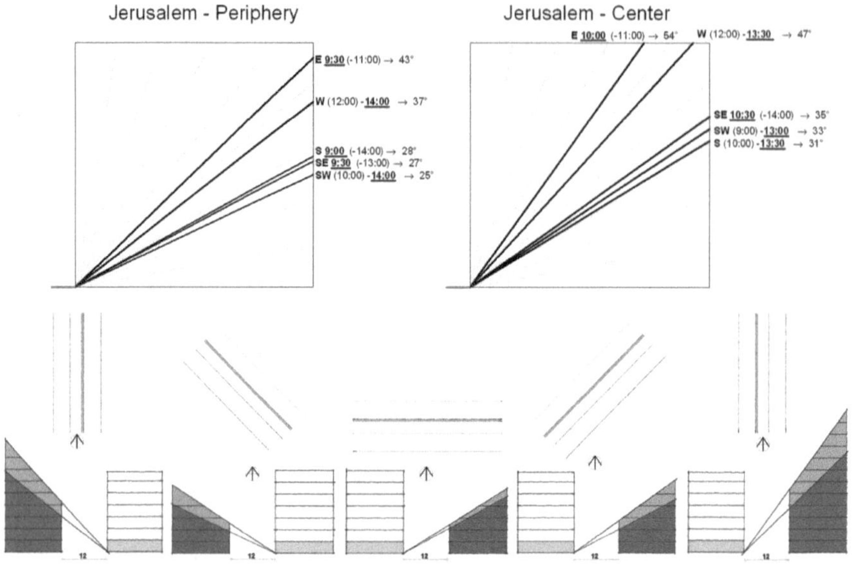

Fig. 4.1 Simple height planning guidelines to keep solar rights in 12 m wide streets in Jerusalem

regulations, which have implications for the early design stage, such as urban planning of street setbacks, tapering of volumes for solar access, or height delimitation (Capeluto et al. 2005). An example of such guides is presented in Fig. 4.1 (top) and their application for typical height guidelines made for 12-m wide streets in Jerusalem, according to different cases (Fig. 4.1, bottom).

These urban guideline tools have been proposed and implemented as master plans and regulations to ensure insolation of sidewalks and public spaces for different cities (Bosselmann et al. 1991; Capeluto et al. 2003; MVRDV 2010; SFPD 2016).

Basic climatic design guidelines are also incorporated on software that evaluates compliance with energy codes. Even though, the estimation algorithms are usually hidden from direct view and users can only input finished alternatives and expect the result as an output. Although these programs only appraise how the projected building is doing compared to the code, designers can learn and improve their projects if that kind of software provides feedback on the points that need enhancement (Shaviv et al. 2005).

4.4 Generation Tools

During the conceptual design phase of buildings and urban areas, the designer deals with several geometrical characteristics. They are related to the building orientation, its height and width, its relationship to open spaces surrounding it or to be planned inside, the use or absence of pedestrian sidewalks, and its relationship with neighboring buildings. Conscious designers are also aware of the fact that new buildings may create a different microclimate from what is currently present (Capeluto 2013). For example, significant height can change the local wind regime, it can also affect daylight access and shading of existing buildings and neighborhoods. These aspects are strongly influenced by early decisions made by the designer.

The determination of a preferable design solution becomes specially complicated due to mutual influences. For example, orientation and proportions dictated by adjacent project streets will influence the exposure of facades and sidewalks to the winter sun, as well as determining the required shading during summer. On the other hand, ignoring solar rights at the preparation stage of a master plan, may cause discomfort conditions around and inside the buildings which cannot be fixed unless expensive measures are taken, seriously compromise the project energy performance.

The early stages of the design process characterize themselves by a constant search of a design direction. Nevertheless, decisions taken in those moments can determine the success or failure of the proposed project (Hari 2001). This was also verified by a study on the design principles of high-performance intelligent facades for hot climates, in terms of both building energy consumption and user comfort (Ochoa and Capeluto 2008).

A relationship between information and design freedom in the different stages of the design process has been described by (Ullman 2009). For Ullman, the amount of design freedom is inversely proportional to the specific knowledge obtained about a given design problem. Both vary at separate rates, according to the relevant design stage taking place. Despite the fact that there is not much information during the early schematic stages of the design process, a high degree of design freedom exists. It is then when major, significant decisions, are made that affect project performance. As the project advances, design decisions made on the basis of information gained reduce the amount of design freedom possible.

Due to these facts, it is important that design tools should support architects from the very beginning of their design process. The ideal tool should provide meaningful information according to the relevant design stage, providing answers to doubts raised by designers such as "what if?", and "which direction to take?". However, at present they are distracted by input issues or the overwhelmingly display of incomprehensible results ("what do these numbers mean?"). Existing design tools are still rudimentary and with many limitations for use by designers in early design stages of any architectural project. They generally are aimed to external consultants, requiring exact data in a stage when designers consider conceptual ideas from a range of options, rather than precise details and numbers. Design tools that suggest solutions based on ideas are still rare (Ochoa and Capeluto 2009).

Generation tools may aid to define certain variables (such as proper geometry) to achieve a given performance level. Performance-driven form generation is used, based on the principle that data can be used to generate architectural form. The idea has been in use for some time, Shaviv (1975) proposed a method and a computerized model for designing fixed external sunshades. The principles of Shaviv's method, when used in a dynamic model, can serve for generation of a changing shading surface according to the conditions of the day and year. The method was extended later for the generation of solar rights envelope for the design of solar communities (Shaviv 1984).

Different researchers developed, mostly between the 70s and 90s, a series of computerized generation tools that calculated specific aspects of the design problem. Many were related to the delimitation of solar rights and access. One example is SustArc, developed by Capeluto and Shaviv (2001). It uses the solar rights envelope (SRE), the solar collection envelope (SCE) and solar volume (SV) data as target functions. These solar envelopes define the space of all possible design solutions that either considers solar insolation or solar shading. SustArc allows the generation of different building configurations, ensuring solar rights of each neighboring building. It can also be extended to the analysis of open spaces such as sidewalks, gardens and squares. The model presents the maximum available volume in which it is possible to build without violating the solar rights of any existing building, as well as the designed one. This type of generative design tools have the potential of providing meaningful information to support design decisions during conceptual design stages. SustArc has been lately implemented in Rhino/Grasshopper and is part of the Ladybug analysis tools suite. In this version it allows the refinement of the conditions for the generation of the Solar Envelopes allowing maximum deployable volume.

Figure 4.2 shows an urban study done with SustArc for the planning of a high-rise business district in Tel Aviv, located next to a low-rise residential area. The tool offered a satisfactory solution that kept the solar rights of the inhabitants (Capeluto et al. 2005), while allowing urban development.

4.5 Generation Tools Incorporated in Existing Programs: SunTools as Case Study

Recent computational developments offer new flexibility in the production of new computer models. No longer is the programmer bound to write a self-standing program, but they can now can be incorporated as complements (plug-ins) to existing programs. This is useful to improve existing design programs, well-known by architects and engineers, without the need of building a new model for the evaluation. In this section, it will be shown how existing tools can be enhanced to provide information with a new meaning for design. It will be done the through the presentation of SunTools (Capeluto 2016).

Implemented using Ruby scripting language as a plug-in for the SketchUp modelling program (SketchUp website 2016), SunTools allows visualization of solar position, sun paths and visible sky. It produces axonometric views from the sun to easily analyze mutual shading, solar access and penetration at any design

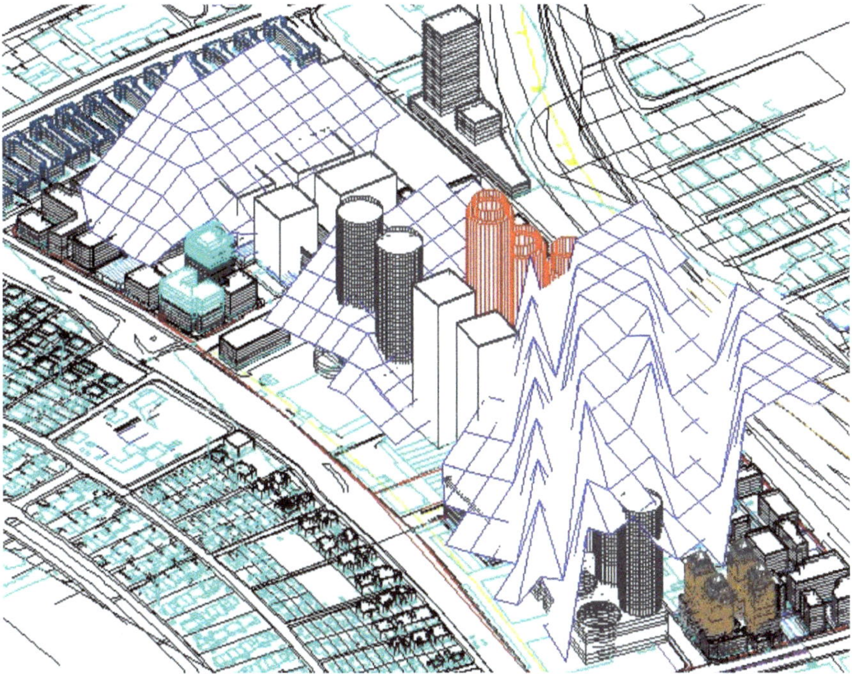

Fig. 4.2 Solar envelope study made with SustArc for planning of a high-rise business district in Tel Aviv, Israel

stage. Evaluation results are provided in a way that can be used as generative information.

One chief advantage of SunTools, is that analysis is done easily within the same program, eliminating the need to export geometric data to external programs and minimizing the learning curve. Therefore, the same 3D schematic model can be used. This tool aims to serve, from the early design stages, a variety of end-users: students, teachers, architects and consultants, in order to include solar consideration in design.

Common modelling tools widely used nowadays by architects, including SketchUp, provide accurate capabilities of shadow casting visualization. However, not all of them allow visualizing the sun as a light source itself or its path. In all cases, the relative position of the sun in the sky is calculated internally according to geographical definitions inside the model.

Visualizing the solar path during a required period of the year, or at a certain date and time, can help to understand better the impact of the sun in relation to the project and its surrounding areas. Since this information already exists in the model, it is a matter of exposing it to the designer, creating a new layer of information that can be worked with (Fig. 4.3).

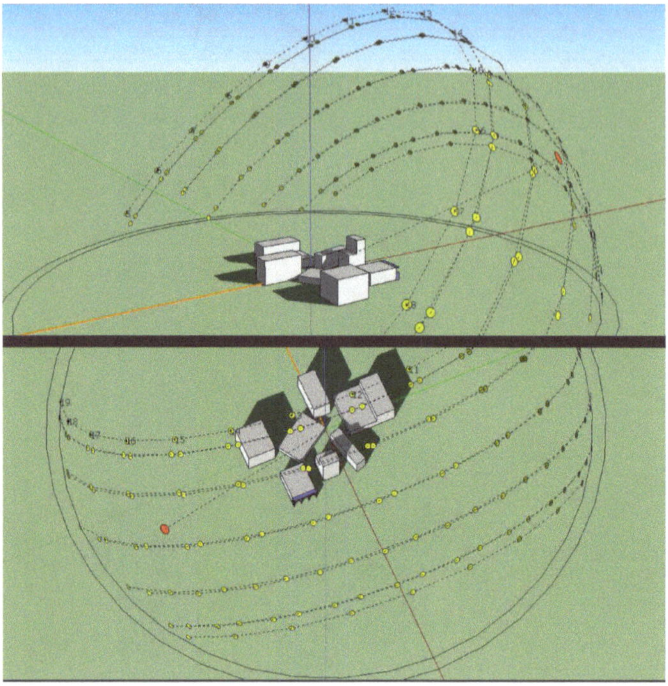

Fig. 4.3 Visualization of the solar path and solar position from the SunTools model

Once solar geometry information has been incorporated as part of the working model, it can be used and manipulated to extend its meaning. This allows performing evaluations of a series of design alternatives that can be interactively modified in order to improve their performance. SunTools uses this information to calculate solar penetration and solar access at any internal or external point in the project. By taking advantage of common capabilities of modeling tools, to produce custom visualizations from pre-set viewpoints and directions, a powerful evaluation of the early stage project can be produced (Fig. 4.4).

The evaluation allows designers to use their own 3D working models, understanding in a single comprehensive view periods of exposure and shading for a given position in the project. Furthermore, designers can see and understand times and causes of comfort problems, modifying the design accordingly, in order to obtain the desired performance (Fig. 4.5).

In a well-designed space, the use of daylight reduces energy costs, enhances visual quality, and provides psychological benefits that cannot be imitated with electrical lighting. Daylighting availability in certain urban areas can be limited due to a densely built environment.

The surrounding built environment can affect daylight use inside buildings, compromising its benefits for end-users of the buildings. Daylight penetration depends on a number of design parameters. Some of these include: room depth

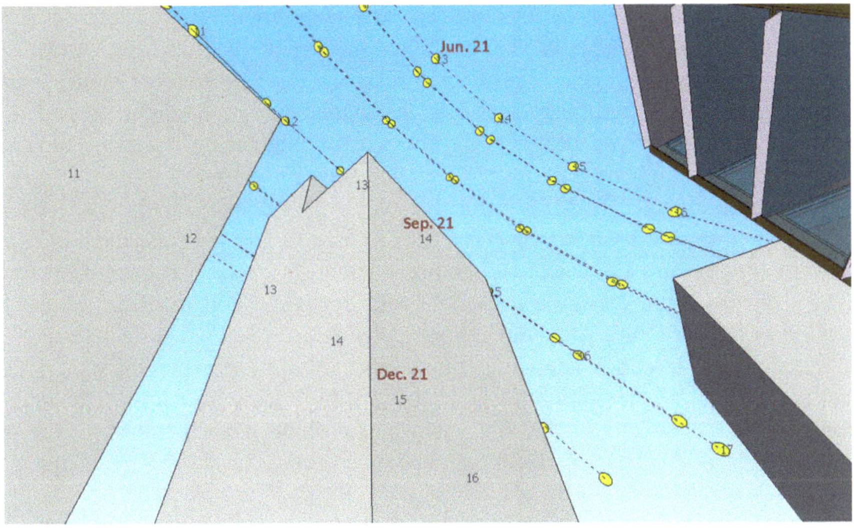

Fig. 4.4 Evaluation of year-long solar access for an open space using SunTools, as seen from the point of view of a hypothetical observer

Fig. 4.5 Evaluation of solar penetration using SunTools for two shade options. Above, a short solar shade, and below, with an enlarged shade to provide protection during required period

measured from the window wall, ceiling height, internal reflectances, window orientation, shape and size, as well as optical properties of the glazing. Adequate parameters of these factors are unknown to designers during the early design stages. However, the most significant factor is the availability of daylight outside the building. This factor can be seriously affected by external obstructions, such as neighboring buildings or significant vegetation such as tall trees.

The sky solid angle (SSA), presents the solid angle subtended by the path of the sky visible from the studied point. The SSA is proposed as an assessment of how external obstructions influence daylighting availability inside buildings (Capeluto 2003). A correlation exists between SSA and daylight factor available (DFave), serving as indicator of a site's daylighting potential.

In SunTools, the SSA can be determined using the 3D model containing volumetric information under study. It is determined by tracing rays from the studied point into all directions to the sky vault, and probing if each ray is visible or obstructed from this position. Both SSA and a percentage of visible and obstructed sky can be calculated.

Results for sky vault blockage can be visualized interactively. Supplementary information can be presented graphically for the designer to use (Fig. 4.6) showing both solar paths and visible sky vault. Similarly, additional information can be gradually added to the model, according to different variables that need to be studied during the design stages. Examples include solar irradiance on the design of building envelope, or the design of open spaces.

4.6 Single-Aspect Evaluation Tools

This section will emphasize the description of tools dedicated to the evaluation, mainly through computer simulation, of a single building physics aspect. They constitute one of the largest portion of software programs available to consultants and researchers (as their main audience), and to advanced designers (as a secondary and minor audience).

Some common characteristics of single-aspect tools

Evaluation tools analyze the performance regarding one or several aspects of a given design alternative. Although the architectural design process provides a single built design, before reaching that single result, in general numerous design alternatives are created and evaluated. Examining several design ideas and alternatives during the design process may be difficult, time consuming and not always possible. As architectural design is often performed under tight schedule and budget, the amount of time and resources designers have to investigate design options is highly limited.

In practice, the large majority of existing evaluation models is geared to simulate and evaluate finished and well defined design alternatives. According to Ochoa and Capeluto (2009) they are generally unsuitable as practical design aids for architects

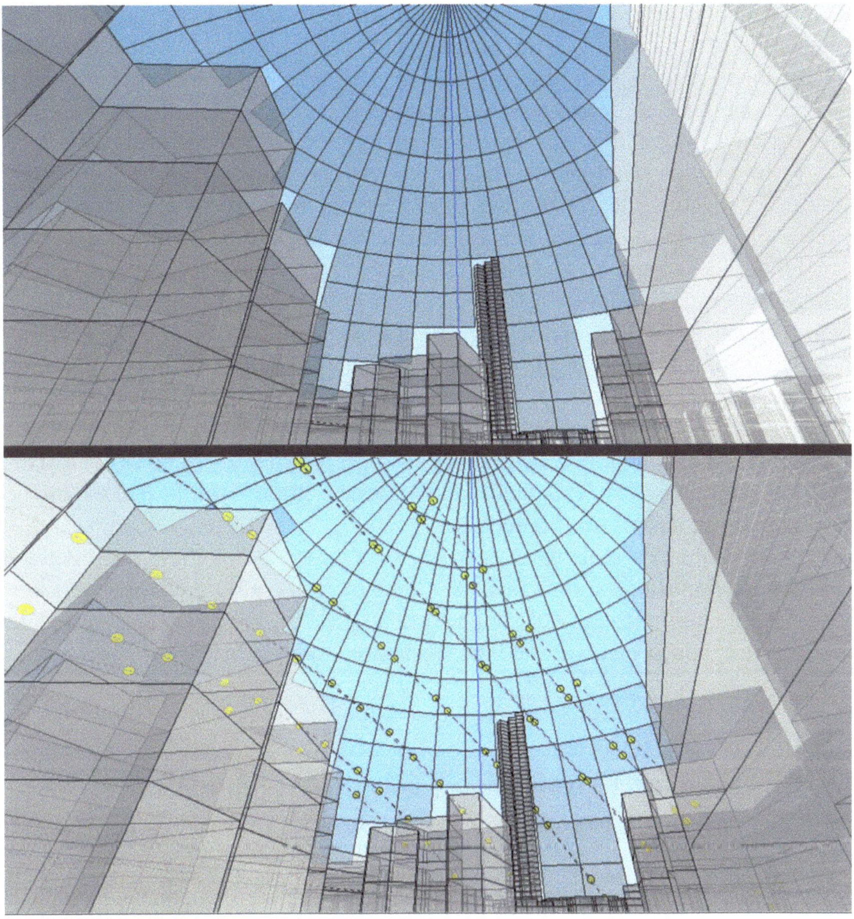

Fig. 4.6 Interactive representation of sky vault and combination with solar paths in SunTools

from the early stages of the design process, since they share the following characteristics:

(a) Most single aspect tools are still designed for use by consultants, who many times are incorporated into the design workflow when the main geometric characteristics of a project have been determined. Using them in this way limits impact on the finished performance.
(b) Many of these programs evaluate and model a certain finished alternative, and cannot suggest nor evaluate a variety of design ideas, options and directions.
(c) It has been demonstrated in previous sections that early design decisions are based on an amount of design freedom inversely proportional to data available then (Ullman 2009). Such elements are hard to evaluate using tools relying on exact data.

(d) The architectural design process logic is not followed by these tools. The main reason is that it involves an iterative, and sometimes loose, method that takes as input changing information, mental design schemes and principles to be met, as stated in design briefs.
(e) Detailed information and precise input is used by current evaluation tools. However, this is not known and many times is not relevant at the beginning of an architectural design process. Architects might not need size of air conditioning equipment, exact window size or glazing combinations, etc., in order to provide a building massing.
(f) Designers can find themselves operating "out of instinct" about which design direction to develop when projects are on the boundary of their expertise. These tools hardly can provide and answer to questions such as "what to do next?"
(g) When using these tools, architects without previous experience on these programs find themselves evaluating detailed, finished alternatives using a trial and error approach. This makes them depend on previous factual experience and not on design exploration.
(h) Many evaluation tools have stand-alone, complex, non-logical user interfaces that require extended time to learn. Complexity and a steep learning curve can make the use of these tools to be concentrated on only a few operators in the practice, which can become a production bottleneck.

4.7 Single Aspect Evaluation Tools: The Lighting Simulation Case

A very good, up-to-date catalogue of programs in this category is maintained by the United States' chapter of the International Building Performance Simulation Association (IBPSA-USA 2016). Their list shows the wide variety of programs that can cover almost every aspect involved in building analysis. Examples range from input calibration to life cycle analysis and commissioning.

Due to the many aspects involved, lighting simulation programs, and their use for designing intelligent envelopes, will be featured as a case study. A detailed survey on these particular programs has been made (Ochoa et al. 2011), which can be consulted for further reading. Where relevant, this section will provide some illustrative examples on the capabilities available to consultants and designers.

Complexity level: Within the same category of simulation software, users can encounter different levels of detail concerning the amount of variables they can manipulate, input required for processing data, type of calculation models and output they can obtain from the tools. Factors that will determine preference for any given software include: existence or not of a graphic user interface (GUI), its reputation within the peer community, acquaintance of users with the subject being modelled by the tool, computer skill level of software operators, overall software requirements (such as hardware or Internet bandwidth), and aims of using the software for a given purpose in the process.

Input data: Data required for initiating calculations is not uniform as well. Although ideally a file format that comes from a previous design stage should be used, most software suites are not able to process them directly. They also might not support importing file formats different from what programmers have stated as usable. Some tools include their own procedures to describe within them the space that will be analyzed. The extended learning curve for adding input, and understanding how the program works, makes them in general to be used by specialist operators.

For lighting simulation, important factors that need detailed consideration in the input are: definition of light sources, definition of existing and projected surface geometries and their properties. Specialized data must be collected beforehand or a safe assumption be made (usually by a consultant).

Two examples can be mentioned to illustrate this point. One is the well-known calculation engine Radiance (LBNL 2014) and DIALux (Dial 2016). Radiance is considered to be the standard for lighting simulation. It is well documented and tested in a variety of situations, for both electric and natural lighting. DIALux, on the other hand, has been known mostly for electric lighting design, even though now it has abilities to model daylight.

However, Radiance has no user interface of its own and commands are mostly written as instructions. Some researchers and casual users have developed their own interfaces for Radiance, but they are not standard. External translators (also written by program users) are employed to import into it scenes and materials created in other programs. DIALux, for its part, has its own drafting interface and can import data made in CAD programs.

Nevertheless, these differences exist since the objectives of each tool are different. Radiance can be used to model and research a wide variety of situations, but DIALux is aimed mainly to artificial lighting designers. Another influence on their usage also comes from their distribution channels. Both tools are free, but Radiance is open-source code, while the one of DIALux is proprietary.

Algorithms and calculations: In terms of procedures used by lighting simulation tools in general, different calculation techniques and theoretical models are applied by each model. Most of them share combinations of forward ray-tracing and radiosity ones. It cannot be said that any given algorithm is superior to the other, but each allows the study of different aspects of the same problem. Particularly for lighting simulation, during its development there have always been tradeoffs between computational speed and accuracy. Although in recent years, with the increase of computer hardware speed, user expectations have also changed.

Detail level: Many researchers are interested in exploring more complex scenes in shorter times, compare several solutions to select the best more than calculating its exact performance, or they want to study physical phenomena that have difficult simulation description and are used for intelligent envelopes. They include among others: analyzing reflectivity over time spans of dynamic elements such as blinds, interaction of natural lighting with electric lighting for energy saving and human usage of shading devices, study of specialized geometries such as laser-cut glazing patterns, and reaction of electrochromic glazing, etc. Success in simulating these elements on commonly available lighting simulation tools depends largely on how innovative or well-known are the element properties. Not all innovative elements can be simulated, and need therefore real-life experimentation, as is the case of materials that have custom-made changing properties according to each situation (e.g. reflectivity of specific light wavelengths for example).

From these analysis types, dynamic simulation is one of the most useful for lighting design within all types of intelligent building envelopes. The challenges posed by changing daylight throughout the day and year bring the need to consider more than single instant calculations. They also define the interaction between daylight systems and electric lighting, which in turn are controlled by a BEMS. Extreme situations that can cause discomfort as very high or contrasting light levels and glare must be identified and the appropriate course of action defined. A good example of a software tool dedicated to this purpose is DIVA (Solemma 2016). The program implements Radiance in a high-end modelling tool. Although this type of integration will be discussed in detail later on, DIVA can perform environmental daylight and energy evaluations of a given variable in a time frame and map them for consultant evaluation.

Output data: Output obtained from these simulation programs is also varied and not always suits the needs of the designer in the specific design stage. It can range from large pure textual files reflecting numerical arrays to visual quantitative or qualitative animations of the results. Numerical data can be post-processed in other programs such as spreadsheets. Of more interest, however, is the production of data that can be processed by other programs that help to continue the design process. This type of output, however, is not widespread.

4.8 Tools for Whole-Building Simulation

Analyzing many aspects of the building envelope with separate tools can be time-consuming. Many times, using different programs will not enable the consultant or designer to appreciate the mutual influence of different environmental variables on envelope and building performance, for example the influence of using daylighting in the overall energy use of the building. To this end, several tools have tackled the integration of these aspects as whole-building simulation programs.

Most of them analyze this integration from the study of energy con-sumption. The analysis of overall energy consumption of a building is highly de-pendent on factors such as thermal properties of the enclosure, existing controls, external weather characteristics and human usage and behavior (among many others). However, integrating the calculation of different factors is not easy, and there is always the interest of providing accurate, practical and meaningful results on a satisfactory time frame. This results in priority being given to one aspect over the other, yet maintaining acceptable results for all the elements being calculated (Ochoa et al. 2012).

User preferences: When evaluating which tool to employ for a given problem of whole-building simulation, users must take into account its accuracy and reliability. It has been demonstrated that more detailed simulation tools have the best simulation performance, primarily in terms of assessing heating and cooling energy consumption. (Yezioro et al. 2008). There are also whole-building simulation tools that have been adapted to cope with the uncertainties given by specialized studies, such as expected energy use in retrofit projects (Lee et al. 2015).

Input methods: Similar to the software programs described in Sect. 4.7, there are different ways to enter input for analysis. Input methods are varied, since project geometry and characteristics of building elements must be entered for the program to understand them. Some programs provide dedicated interfaces, which help to organize numeric or descriptive input, through a series of menus and areas to enter required data. Others have a mixed approach, where geometry is input directly or imported from a CAD file, while data such as U-values or schedules are input numerically (Alliance for Sustainable Energy 2016a, b).

Calculation methods and representation level: It is also of interest to note that some of these energy calculation programs can collect input, sending it to a *calculation engine*—a program that has no interface of its own but receives and provides results, according to commands being sent to it. Examples of this instance are DesignBuilder (DBS 2016) and ENERGYui (Eui 2016) which use EnergyPlus (DOE 2016) as their calculation engine.

Regardless of the data entry method, or where it will be processed, prior training and understanding of building physics and its usage are required to provide the program with correct data. This can restrict the number of users for these tools. Some programs, such as eQuest, strive to make their interfaces more intuitive for all design team members to use (Hirsch 2014).

Most of the leading calculation engines work with "objects", modular programming elements that are used to describe one aspect of the simulation, such as wall or glazing properties according to different variables such as U-value, material layers, reflectivity, etc. In most cases, the structure of these objects is fixed for the regular user, but can be customized and extended through manipulation of different files (EnergyPlus) or through custom interfaces for component modification (TRNSYS 2016).

The most complete software tools can model the energy contributions from a wide array of current and innovative elements. The current approach used is to model dynamic simulation through a time range, usually a year or a season, in order to study the reaction of these elements. For example, users can ask the performance of photovoltaic elements throughout the year, the cumulative solar radiation on a given wall that will be covered with phase-change material, or the influence that automated blinds have over the use of cooling and heating systems or occurrence of glare.

Simulation of control strategies for intelligent envelopes get a closer representation of reality in this type of programs. Some of them have specific programming objects that are used to represent a building energy management system. They also require a realistic spatial location of sensors, upon which systems such as lighting and blinds will be deployed.

Concerning the representation of highly customized or experimental elements, this can have some degree of difficulty. Accurate performance data, translated to physical formulas, can only be obtained from thorough testing of the element prototype in different situations.

Output data: Output provided by whole-building simulation tools can be provided in formats that allow post processing in other programs for data visualization or data analysis. As such, most of it is numerical and requires specialized knowledge to interpret and understand it correctly. Acquaintance with expected building performance is needed to understand results, or to debug the program in order to fix any errors that interfere with correct calculations. This makes the process tedious, as there is no guidance provided within the programs themselves for this purpose, or to determine what to change in order to improve the design. Tools that have a visual input usually provide a built-in visual display of simulation results in the form of graphs or renders.

Specialized plug-ins and software can also display visually energy consumption as an effect of dynamic factors (Trimble 2016a). Export capabilities of these tools to CAD or BIM formats are still somewhat limited, due to the different requirements of each specialization.

4.9 Other Tools

In addition to the programs and cases mentioned in the previous sections, a wide variety of "hybrid" programs exist, in which their primary purpose is not energy simulation but they are complemented by it. Nevertheless, they can represent the future of design analysis through the integration of energy aspects with building modelling.

Examples of the above can be found in different CAD and BIM products. Inside them solar studies are offered as easy-to-use built-in features (Trimble 2016b), and where lighting and energy analysis can be performed directly on the model (Autodesk 2015a). The importance that this represents lies not in the accuracy of the algorithm being used, but in the integration being offered of environmental analysis with a drafting and modelling tool frequently used by architects and designers. Any change in the building envelope can be evaluated at once, without requiring any export to a third-party program or sending the file for analysis to a consultant. However, at present these programs still fall behind in the type of guidance provided to the BIM operator for feedback.

CAD and BIM programs are also now seen as the platform upon which researchers can offer plug-ins for specialized analysis, or where they can couple advanced energy simulation suites. One example of this is OpenStudio (Alliance for Sustainable Energy 2016a, b). It is offered as a plug-in that interacts with a

SketchUp model while allowing to introduce on the interface input that will be calculated through EnergyPlus. Other propositions will extend the capabilities of BIM-based programs to further stages of the design, including code compliance (Cheng 2014).

Whole-building simulation tools are also proposed for the operation of the intelligent building itself. The assumption is that advanced forecast information can help preparing different elements (within the envelope, for example) to respond better to localized weather events, such as variable temperatures during the transition seasons (Chou and Bui 2014).

Nevertheless, one area where tools in general are trailing is in the smart selection of alternatives and guidance to the user in the early design stages. Although advanced work has been done for smart selection systems in other fields such as medicine, this is still not the case for building design. However, certain examples exist, even if at the demonstration stage, such as NewFacades (Ochoa and Capeluto 2009; Petersen and Svendsen 2010).

Analytical tools like Honeybee and DIVA also offer an insight into future directions of simulation programs, where parametric simulation and result analysis can be presented in an enhanced way. One of the strengths of these programs is the visualization of dynamic simulation results. They are of particular interest to represent the interaction of intelligent envelopes when applying different strategies for comfort and energy savings combined.

A nascent category for analysis tools is that of web-based applications. Although having their roots in the early years of the 21st century, at the moment these web tools have limited usage. Factors affecting them include technical requirements of coupling programs that have a linear processing nature, with those that can handle multiple requests at the same time. Another limitation that has bogged down their development is the lack of uniform bandwidth speed access from the user-side. An increased internet access speed would allow an acceptable experience, similar to that of the desktop program. The most used approach in many of these tools is that of consulting pre-defined databases. Nevertheless, real time calculations offered online are gaining ground.

Some of the first web-based tools were geared towards diagnostic of an existing situation, providing some advice (LBNL 2016). They have evolved towards optimization and decision support (Xylem Technologies 2016). Cloud computing integrating whole-building calculation engines can also be found (Greenspace Live 2016; Autodesk 2015b).

4.10 Design Suggestion Tools

As mentioned before, design suggestion tools that follow the logic of the architectural process are rare. To be successful, such tools need to take a programmatic approach, where many variables are assumed to have an acceptable numeric value (a smart default) according to the initial situation described by the designer (Ochoa and Capeluto 2009). Preferable *design directions* may be suggested to the designer through the analysis of the results of a series of simulations generated by the system regarding certain design variable, for example optimum windows size. Designs should also be generated according to a set of climatic principles. Some of these principles can be stated as "pseudo instructions" of conditional logic that the program has to follow, filtering possibilities. An illustrative (albeit very global) example is the next sentence:

If the project is located in a hot climate, ***Then*** always consider shading elements

Further refining of the logical sentence would ask which type of elements, where they wish to be located in the façade, their dimensions and properties, etc.

Although conditional logic or Boolean algebra are some of the pathways to obtaining solutions, they provide general principles for developing such programs. Database lookup also provides another alternative, but the elaboration of the database itself can be time-consuming and must be flexible for future growth.

An adequate query system must also be present, which must be tailored to typical design questions being made during the early design stages. It is a non-trivial task that can determine if the program will be relevant or not. Issues such as wording, questionnaire sequence and length, are among items that need to be considered.

4.11 NewFacades: Advice Tool for Early Design Stages of Intelligent Envelopes

This section presents an experimental design suggestion tool named NewFacades (Ochoa and Capeluto 2009). As such, it is intended to assist architects in the early design stages of intelligent envelopes. It was written as a standalone tool that provided design directions and characteristics, which in principle help comply with different energy saving certifications. For the testing phase it complies with the prescriptive approach of the Israel energy rating standard for office buildings (Shaviv et al. 2008). By using elements from the prescriptive approach, the tool focuses on suggesting a variety of near-optimal possibilities, instead of optimizing them. Suggesting multiple and diverse good initial alternatives is more beneficial for the early design stages when designers explore various design routes.

The general approach used by the tool, is of gradually exposing the architect to description of details composing the final product as the design advances. Exact numeric properties of any building element and systems are irrelevant for use in early design stages. Yet, they are important when interacting with building system experts, who will perform accurate simulations. The presented tool uses features from expert and rules-based systems, interpreting input data that is known at the initial design stages. Final concepts are stated similarly to a catalogue, shown together with a comparative outlook of building energy consumption and visual comfort for each option.

The tool uses a predefined components library based on descriptions that can be used by the whole-building simulation program EnergyPlus, whose installation is a pre-requisite, but not mandatory to operate the program. It was chosen due to its open-format nature and that it is widely known by the building physics community.

These descriptions are combined according to input interpretation, and then evaluated by the program. Presets for many of these descriptions, such as air change or infiltration rates, comply with the Building Energy Rating System recommendations. These presets can be consulted in the element library inside the program. However, they are not revealed to users on initial program runs as they are not usually known in early design stages.

Design components in the library can be divided into "geometrical design parameters" and "non-geometrical design parameters". The first category comprises elements that architects manipulate. They are also part of the search objectives in early design stages, such as orientation, building massing and morphology, window size and location, need for shading devices, etc. The second category includes elements influencing physical simulation, but whose exact determination in early stages would deviate and slow down the design process. Examples include insulation details and values, infiltration rates, details for HVAC equipment, etc. These parameters are modified according to the needs posed by each design step.

By asking relevant questions to a given situation, presets are consulted in the library, all design components that have code compliance within the desired design direction are selected, and the entire files are sent for evaluation by the whole-building simulation program.

This process is able to generate multiple options, as many as allowed by the variety of elements within the library. Each alternative is generated as a separate input file that can be evaluated by EnergyPlus. In this way, any option that the program generated can be modified later, using other programs that read those input files. The designer can also transfer these files to consultants for evaluation.

Concerning input, the main user interface for the program is shown in Fig. 4.7. The interface is divided into four working areas: input collection, design principles (corresponding to the performance certification chosen), processing buttons that allow analysis with or without EnergyPlus, and output viewing and manipulation. The layout presents itself in a simplified way, reducing time spent by the user learning the tool itself.

The logic for building design alternatives follows the application of six main rules (Fig. 4.8). Three of them infer design intentions, and the remaining three determine the type of behavior that facade elements, filtered by the first set of rules, must have according to a chosen degree of technological complexity. Therefore, the first set of rules needs user input, and the remaining three complement them. According to the chosen rule, the program consults which relevant elements inside the library comply with them.

A pre-defined basecase is added to the set of design alternatives. It gives an idea how intelligent envelope elements improve the design compared to a standard

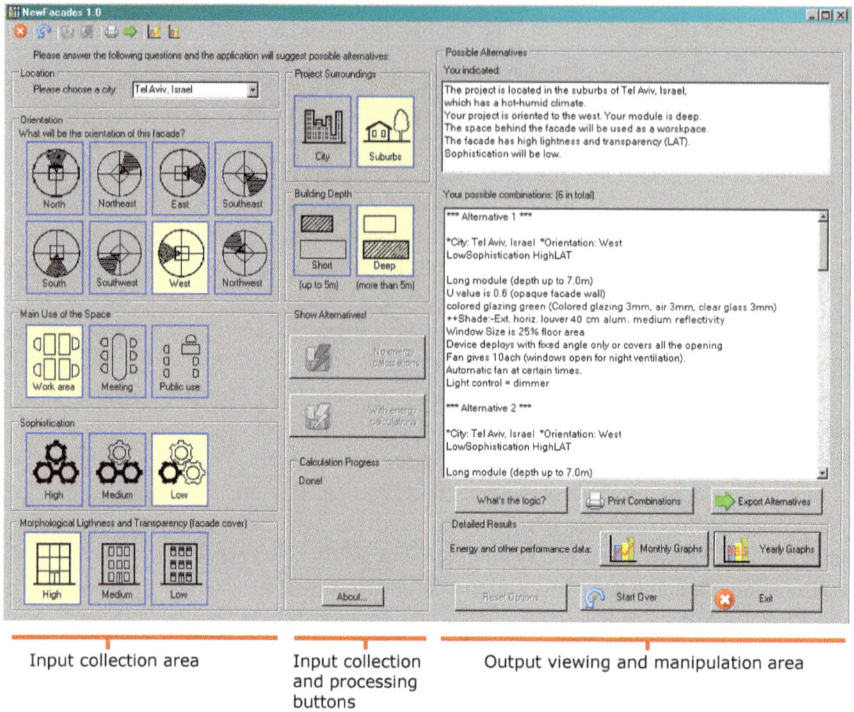

Fig. 4.7 Main interface of NewFacades with outline of data areas

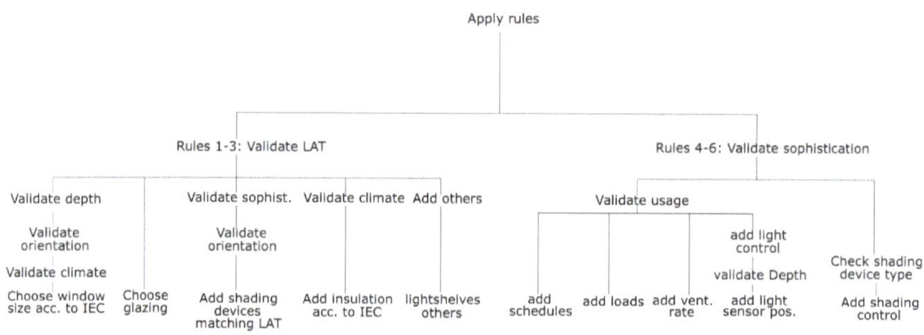

Fig. 4.8 Flow of design rules in NewFacades. (*LAT* Lightness and transparency)

reference case. This reference case is found in most energy certifications (reference building).

In order to introduce diversity in the solutions that are suggested, and to avoid bolting the designer in a single direction, the tool chooses a given number of random facade items that are different to what the user specified. For example, if a clean external façade was desired, then it can add some options with external devices. New combinations are provided in this way every time the program is run, even with the same initial conditions, presenting the designer with diverse results and alternatives.

The energy simulation engine writes, in a previously specified folder, its own result files for each of the alternatives that are generated. Each file has de-tailed information in them about the performance of each alternative. However only energy consumption and daylight glare index (DGI) are shown as performance indicators. For energy, heating, cooling, lighting and ventilation are provided. These results are shown to the user in a comparative way through graphical means (Fig. 4.9). Indications are also provided, in the case of visual comfort, of the periods when the alternative does not comply with the upper DGI limit. It can serve as indication for the addition of other remedial measures.

There are options to view more detailed results such yearly totals or to focus on monthly ones. The program also presents, in a text format file, a detailed list of all the alternatives that were generated according to the input and design rules used. A sample result table is provided in Chap. 5, Table 5.2.

Based on the results provided by the program, designers might consider refining the initial scheme, according to what is allowed by project conditions. This information is known to the designer but can be translated to program input. Simpler changes would suggest experimenting with other external/internal devices, or increase technological complexity of the actuators. Changes to geometrical factors, which would depend on specific project requirements, include exploring

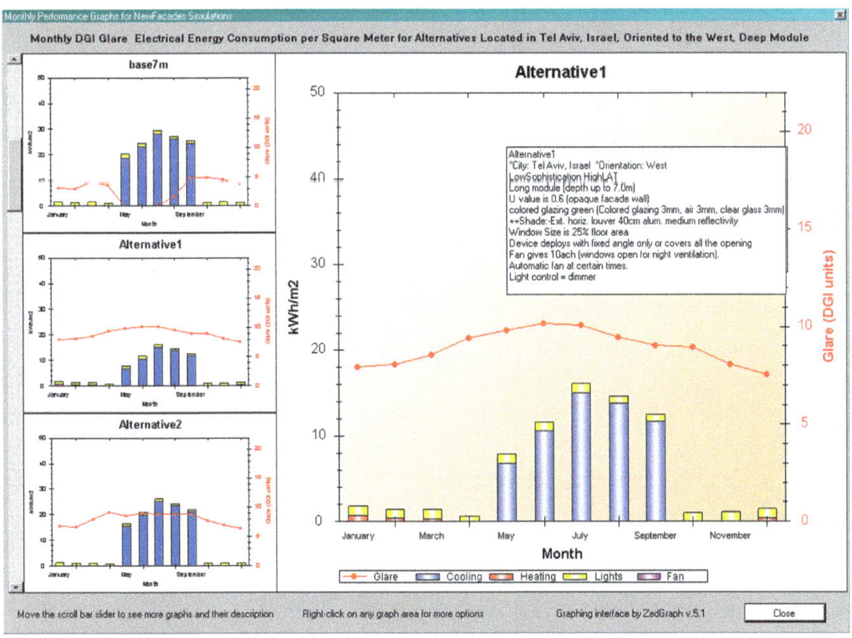

Fig. 4.9 Sample graphical output given by NewFacades

changes in orientation or the area of openings. These prospects can be verified with the tool.

The designer obtains through visual output a simple, quick and comparative assessment. It aids to consider which direction to take for additional exploration, and to understand the performance implications of certain design decisions. It allows architects to choose alternatives based on design criteria, eliminating guesswork regarding code compliance and visual comfort.

As mentioned previously, NewFacades also generates as part of its output complete EnergyPlus input files. Such feature allows further analysis and development of the design alternatives, without the need to translate any other type of files. In this way, small pauses in the design process caused by file compatibility are eliminated and the workflow can become more continuous.

A designer using a tool such as NewFacades can pass from idea to architectural concept, since the design rules are based on climatic and visual comfort strategies complying with an energy certification or standard. As a series of intelligent envelope alternatives are generated and evaluated for different aspects, designers can have an improved assessment when selecting a given design direction for further exploration. The selection becomes dependent on other factors that are not energy-related, for example budget or technical feasibility to implement a given actuator.

Since NewFacades was developed as a proof of concept, other capabilities that can be added in the short term include: allowing users to include custom locations and weather files, add new building usages and technological systems, performance metrics, and output types. A more advanced implementation would allow fine-tuning the simulation engine from the tool interface itself for custom operation.

NewFacades represents a starting point to a wider study on design guidelines of high-performance intelligent building envelopes, located in a variety of climates. Further developments may include taking one or more of the generated alternatives, and make changes for any of its components understanding the consequences of these changes in real-time.

4.12 Conclusions About the Tools

This chapter has shown the different types of tools available that designers and consultants can use for the development of intelligent envelopes, starting from the early design stages to evaluation of different alternatives. An important characteristic, shared by a large number of them, is the ability to dynamically model the behavior of components throughout a typical simulation year. Such ability allows to accurately reflect and predict how intelligent envelopes will respond to changing weather conditions. It also supports the study of different combinations in order to find the most appropriate one for each situation.

These tools provide the advantage to *simulate* performance characteristics of a constructive element or an entire building before it is constructed. Different alternatives can be considered and evaluated in the form of *scenarios,* which goes beyond a single numeric answer. This helps to identify possible weaknesses in the proposed design and decide on corrective actions during the design phase. The feature is particularly helpful, testing performance at extreme conditions that envelope elements can sustain.

Using computer-based tools in the design of intelligent envelopes and buildings has seen advantageous developments. Current computer tools exploit the abilities of modern computer processors, allowing the completion of multiple complex calculations in very short time frames.

It is foreseen that in the future, there will be a mature development of *internet-based calculation tools*, where the calculation engine resides in a server away from the user's terminal. This method can find adepts, only if the promised calculation complexity is done seamlessly and solved faster than what can be done with desktop-based programs, or if the program has access to a large, heavily updated database.

Input, either for computer or internet-based tools, requires improvements in terms of *streamlining it with the design process* and to *guide the user.* Potential errors and guesswork can then be minimized, both at the start and end of the simulation process. Complex data input, and data translation from different file formats are still found due to the nature of traditional programming. Exact numbers are required to define a variable as dictated by a given computer language. It has been shown that designers work with ideas rather than concrete numbers. Thus, any computer program using exact variables is mostly unsuitable for the early design stage. In addition, many of these tools are geared towards consultants and researchers, who are interested in developing one aspect of the design performance requirement, while designers have to oversee all aspects in almost real time.

Design tools can become even more useful if they offer files for direct use by other programs, to *continue the design process*. However, currently most of the tools today offer basic output files that can be post processed by other programs (for producing graphs, for example) or visualized within the software itself only. File formats given as output by the most suitable tools allows *collaborative sharing* of results across the practice or with consultants in both directions, being useful for interaction between different stakeholders in the process.

References

Agafonkin V (2016) SunCalc. Available at http://suncalc.net. Accessed Mar 2016

Alliance for Sustainable Energy (2016a) OpenStudio software. Available at: https://www.openstudio.net/. Accessed May 2016

Alliance for Sustainable Energy (2016b) OpenStudio 1.11. Available at https://www.openstudio.net/. Accessed Apr 2016

ASHRAE (2016) ASHRAE psychrometric chart app. Available at https://www.ashrae.org/resources–publications/bookstore/psychrometrics. Accessed Mar 2016

Autodesk Inc (2015a) Create a lighting solar study. Available at https://knowledge.autodesk.com/support/revit-products. Accessed Mar 2016

Autodesk Inc (2015b) Insight 360. Available at https://insight360.autodesk.com. Accessed Apr 2016

Bosselmann P, Arens E, Dunker K, Wright R (1991) Sun, wind, and pedestrian comfort. A study of Toronto's Central Area. The Department of Planning and Development, City of Toronto

Capeluto IG (2003) The influence of the urban environment on the availability of daylighting in office buildings in Israel. Build Environ 38:745–752

Capeluto IG (2013) Design tools for asking questions: evaluation vs. generation. EAAE Trans Archit Educ 61:602–613

Capeluto IG (2016) SunTools v2.4 SketchUp plug-in. Available at http://gcapeluto.technion.ac.il/SunTools/STdownload.html. Accessed Mar 2016

Capeluto IG, Shaviv E (2001) On the use of the solar volume for determining the urban fabric. Sol Energy 70:275–280

Capeluto IG, Yezioro A, Shaviv E (2003) Climatic aspects in urban design—a case study. Build Environ 38:827–835

Capeluto IG, Yezioro A, Bleiber T, Shaviv E (2005) From computer models to simple design tools: Solar rights in the design of urban streets. In: 9th International IBPSA Conference, Montreal, Canada, August 15–18

Cheng JCP (2014) A bim-based web service framework for green building energy simulation and code checking. J Inf Technol Constr 19:150–168

Chou JS, Bui DK (2014) Modeling heating and cooling loads by artificial intelligence for energy-efficient building design. Energy Build 82:437–446

CLEAR, Comfortable Low Energy Architecture (2016) LEARN package. Available at http://new-learn.info/packages/clear/thermal/climate/sun/sunpath_diagrams.html. Accessed Mar 2016

DBS—DesignBuilder Software Ltd (2016) Design Builder v4. Available at http://www.designbuilder.co.uk/. Accessed Apr 2016

DIAL (2016) DIALux software. Available at https://www.dial.de. Accessed Apr 2016

DOE—United States Department of Energy (2016) EnergyPlus software. Available at https://EnergyPlus.net/. Accessed Apr 2016

EUI—Energy UI software (2016) ENERGYui. Available at http://energyui.com/. Accessed July 2016

Givoni B (1998) Climate considerations in building and urban design. Van Nostrand Reinhold, New York

Greenspace Live Ltd (2016) gEnergy cloud computing. Available at https://greenspacelive.com. Accessed Mar 2016

Hari A (2001) Integrated, customer driven, conceptual design methodology. Dissertation, Technion Israel Institute of Technology

Hirsch JJ (2014) eQuest v3.65. Available at http://www.doe2.com/equest/. Accessed Mar 2016

IBPSA USA Chapter (2016) Building energy software tools directory. Available at http://www.buildingenergysoftwaretools.com/. Accessed Mar 2016

Kelsey J (2015) kW psychrometric functions. Available at http://kw-engineering.com/resources/software/psychrometrics.php. Accessed Mar 2016

LBNL-Lawrence Berkeley National Laboratory (2014) Radiance v4.2. Available at http://radsite.lbl.gov/Radiance/framed.html. Accessed Apr 2016

LBNL-Lawrence Berkeley National Laboratory (2016) Home energy saver. Available at http://homeenergysaver.lbl.gov/. Accessed Apr 2016

Lee SH, Hong T, Piette MA, Taylor-Lange SC (2015) Energy retrofit analysis toolkits for commercial buildings: a review. Energy 89:1087–1110

Milne M (2016) Climate Consultant v6.0. Available at http://www.energy-design-tools.aud.ucla.edu/. Accessed Mar 2016

MVRDV (2010) Bastide-Niel master project. https://www.mvrdv.nl/projects/zac-bastide-niel. Accessed July 2016

Ochoa CE, Capeluto IG (2008) Strategic decision-making for intelligent buildings: comparative impact of passive design strategies and active features in a hot climate. Build Environ 43:1829–1839

Ochoa CE, Capeluto IG (2009) Advice tool for early design stages of intelligent facades based on energy and visual comfort approach. Energy Build 41:480–488

Ochoa CE, Aries MBC, Hensen JLM (2011) State of the art in lighting simulation for building science: a literature review. J Build Perform Simul 5:209–233

Ochoa CE, Aries MBC, van Loenen EJ, Hensen JLM (2012) Considerations on design optimization criteria for windows providing low energy consumption and high visual comfort. Appl Energy 95:238–245

Olgyay V (1963) Design with climate: bioclimatic approach to architectural regionalism, 4th edn. Princeton University Press, Princeton

Petersen S, Svendsen S (2010) Method and simulation program informed decisions in the early stages of building design. Energy Build 42:1113–1119

Prinsloo G, Dobson R (2014) Solar tracking. Prinsloo-Dobson, South Africa. Electronic document ISBN 9780620615761

Roudsari MS (2016) Ladybug+Honeybee. Available at http://www.grasshopper3d.com/group/ladybug. Accessed Mar 2016

SFPD San Francisco Planning Department (2016) Find My Zoning website. http://sf-planning.org/find-my-zoning. Accessed May 2016

Shaviv E (1975) A method for the design of fixed external sunshades. Build Int 8:121–150

Shaviv E (1984) Design tools for solar rights and sun-shades determination. In: Proceedings of the 9th National Passive Solar Conference, ASES, Boulder, Colorado

Shaviv E, Yezioro A, Capeluto IG (2005) A prescription method for the design of office buildings: an energetic-economic approach. In: 9th International IBPSA Conference, Montreal, Canada, August 15–18

Shaviv E, Yezioro A, Capeluto IG (2008) Energy code for office buildings in Israel. Renew Energy 33:99–104

SketchUp (2016) SketchUp website. https://www.sketchup.com/. Accessed May 2016

Solar Pathfinder (2016) Solar pathfinder software. Available at http://www.solarpathfinder.com/. Accessed Mar 2016

Solemma LLC (2016) Diva for Rhino. Available at http://diva4rhino.com/. Accessed Apr 2016

Szokolay SV (2007) Solar geometry, 2nd edn. PLEA: Passive and Low Energy Architecture International and Department of Architecture, Queensland University, Brisbane

Trimble Navigation Ltd (2016a) Sefaira Architecture plug-in. Available at http://sefaira.com/. Accessed Mar 2016

Trimble Navigation Ltd (2016b) SketchUp. Available at https://www.sketchup.com. Accessed Mar 2016

TRNSYS (2016) TRNSYS—Transient System Simulation Tool. Available at http://www.trnsys.com/. Accessed Apr 2016

Ullman DG (2009) The mechanical design process, 4th edn. McGraw Hill, New York

UO SRML, University of Oregon Solar Radiation Monitoring Laboratory (2016) Sun Chart program. Available at http://solardat.uoregon.edu/SunChartProgram.html. Accessed Mar 2016

Xylem Technologies Ltd (2016) Semergy. Available at https://www.xylem-technologies.com. Accessed Apr 2016

Yezioro A (2009) A knowledge based CAAD system for passive solar architecture. Renew Energy 34:769–779

Yezioro A, Shaviv E (1996) A knowledge based CAAD system for determining thermal comfort design strategies. Renew Energy J 8:133–138

Yezioro A, Dong B, Leite F (2008) An applied artificial intelligence approach towards assessing building performance simulation tools. Energy Build 40:612–620

Chapter 5
Application Examples

5.1 Scenario One: New Office Building

In this section, we exemplify the main steps that would be followed when selecting components for a new building incorporating intelligent envelope elements. For this case, it is assumed that construction projects in that area are not bound by legal limitations that could be imposed on preservation and technological choices. In this way, it can be illustrated how energy and economic constraints are used to determine the space of solutions. We also assume that the consultancy team works closely with the leading architectural team the, both having clear ideas and aims about the climatic principles involved in the design of an intelligent building.

In this first scenario, the hypothetical building is located in the suburbs of Tel Aviv (geographic coordinates: 32° 4′ N 34° 47′ E). The general topography of the area is relatively flat, weather conditions have high influence from the nearby coastline. Local climate has been studied and recorded for an extended period. It is described as a hot-humid Mediterranean one (Köppen-Geiger classification *Csa*). Winters are generally mild and wet, while summers are humid with no precipitation. Although averages for the main seasons might seem moderate (18 °C for winter and 30 °C for summer), the city can experience extreme temperatures and heat stress.

For winter, temperatures can reach below freezing point in record cases. In summer, extremes nearing 40 °C have been recorded, as well as heat stress events that occur during the transition seasons (March, April, November). Precipitation is distributed mainly between the months October to May, and can amount to 528 mm annually. Precipitation can also vary between record years of high rainfall or extreme drought (Bitan and Rubin 2000). Multi-annual climatic data for this location is also available electronically in typical meteorological year (TMY) format, for use in computer simulation programs that use it as part of their inputs.

G. Capeluto and C.E. Ochoa, *Intelligent Envelopes for High-Performance Buildings*, Green Energy and Technology, DOI 10.1007/978-3-319-39255-4_5

The project under study consists of a medium height structure, to be used as a high-tech office building. With an open plan configuration, there will be different unknown tenants occupying several floors. Some office space will be left for speculative occupancy in its different levels. These facts, and the high climatic variation that takes place during the year, makes it suitable to apply an intelligent adjustable envelope system. It will be used to manage different thermal and usage requirements, which are expected to have a high degree of variation inside and outside the building.

(a) **Study of site and context to find alternatives**

In addition to the weather characterization presented in the previous section, site analysis would be made by the design team. In our hypothetical example, the site allows direct views of the Mediterranean Sea, which is located to the west. These views are preferred by the owner as a strong selling point for the project. There are also sea breezes coming from that direction, having the potential of being used for natural ventilation.

The surrounding built environment can seriously affect the possibility of using daylighting inside offices. In medium and high density zones of the city, where office buildings are generally located, the lack of light from the sky at street level can cause design problems. Architects wishing to use daylight to provide a high quality working environment, and as an energy efficient design strategy, must consider these kinds of conditions. Given that only the upper floor in multi-story buildings can eventually make use of skylights, the most common source of daylighting inside the office space is through side windows. Additionally, providing side-daylit offices places limitations on building depth and interior organization (Capeluto 2003).

To simplify our analysis, we will consider in this example that there is no significant shading from any neighboring buildings in the four main orientations. Solar radiation reflected from the ground surface, which can be a source of glare, should be also considered.

The architectural team, for their part, has expressed that the building's image has to convey prestige, technology and innovation. As noted above, West sea-facing views are preferred by the owner for the main building orientation as the selling point to the project. Therefore, window area is expected to be maximized as possible (within the limits of acceptable climatic design and visual comfort).

The feasibility of integrating energy self-generating technologies (e.g. photovoltaic panels) should be studied, considering incentives for the sector, and payback period beyond what is desired by the investors. Intelligence as a design element can be used to project a sign of novelty, help reducing energy use and emissions and ensure occupant comfort. The type of desired tenancy regime calls for the installation of a centralized BMS with enough local overrides.

With such data and ideas, an initial climate analysis can be made to delimit a series of possibilities for intelligent envelope elements. One type of initial examination would include finding the most appropriate design strategies, based on the psychrometric chart for bioclimatic design. Climate analysis using specialized software for the location is possible, given the existence of electronic yearly data. In case this data is not available, specialized procedures would need to be followed. These may include the production of synthetic weather data, based on calculations for the closest location to that being studied (Meteonorm 2016).

The psychrometric analysis for Tel Aviv, performed with the software Climate Consultant (Milne 2016), would reveal a series of general strategies as illustrated in Fig. 5.1. It shows that comfort levels are only achieved on a rather reduced number of hours for that particular climate. Therefore, it is necessary to employ passive or active climate design strategies to extend comfort times.

In this case, in order to extend as much as possible the comfort period reducing the use of mechanical means, the following strategies are suggested by the analysis:

For the summer period:

- Sun shading of windows
- Forced day or night ventilation
- Supplemental mechanical cooling for extreme peaks.

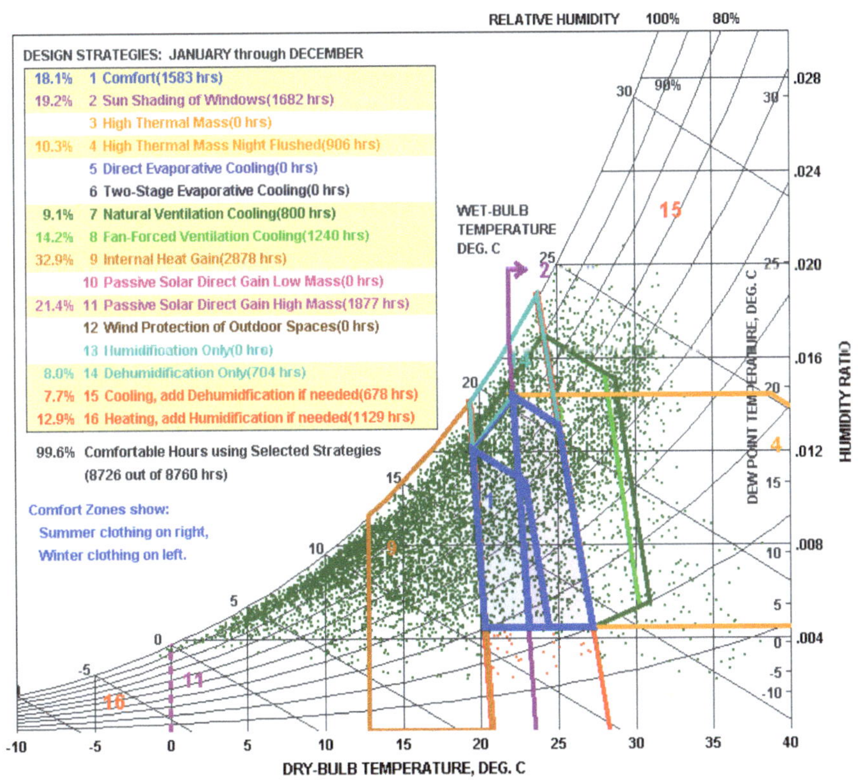

Fig. 5.1 Psychrometric analysis for Tel Aviv, Israel. Performed using Climate Consultant v6.0 (Milne 2016)

For the winter period:

- Maintaining and reusing internal heat gains
- Passive solar heat gain
- Supplemental artificial heating for extreme peaks.

A properly oriented building can maximize the effect of any strategy, enhancing the performance of particular envelope elements. Based on climatic principles, the best way would be to mass the building on a North-South main orientation. Nevertheless, for the current example a starting point as required by building's owner of West-East orientations will be studied to show the different consequences of the decision. The advantages of a North-South orientation considering energy and comfort conditions will be studied and presented as well.

When selecting concrete actuators for meeting the strategies, they can be met in large measure by different families of intelligent envelope elements, as demonstrated in Chap. 2. A sample choice of actuator elements that can be considered by the design team is shown in Table 5.1.

In the table, lighting is included as a strategy even if this was not mentioned as such in the psychrometric analysis. In a typical office building in Israel, lighting accounts for about 30 % of the total electricity load (Ochoa and Capeluto 2006). Moreover, heat gain due to electric lighting can account for a significant part of the total building cooling load during the hot season. Daylight can be used to reduce lighting energy use and heat gains associated with electric lighting. Efficient

Table 5.1 Sample choice actuator elements

Strategy	Strategy detail	Family elements	Example
Cooling strategies	Sun shading of windows	Blinds	Internal, external → define an activation strategy
		Shades	Rolling shades
		Glazing	Low transmittance glazing, electrochromic glazing
	Ventilation	Mechanical ventilation	Electric fans
		Natural (day/night) ventilation	Operable windows
		Air intakes	Building design
Heating strategies	Reuse and keeping internal heat gains	Insulation	Conventional insulation panels
		Solar buoyancy	Active, passive
	Passive solar heat gains	Orientation	Adequate orientation and heavy materials
		High latent heat storage materials	PCM, insulation
Lighting strategies	Lighting	Lighting controls	Dimmer, on/off
		High-efficiency light bulbs	LED, fluorescent

utilization of daylighting can dramatically reduce total electricity load and peak demand, additionally contributing to the well-being and productivity of workers.

Based on these facts, it can be stated that proper illumination of interior spaces using daylighting is also an important factor that must be considered in the planning of an intelligent building. This is a reminder that a complete preliminary study has to go beyond the envelope, and consider all elements involved towards efficient performance and enhanced human comfort.

Previous economic analysis and market availability for specific actuators can be used to help filter which alternatives are viable for construction. Consequently, the choices that will be studied under whole-building and specific simulation programs can be narrowed.

(b) **Evaluation of alternatives**

The number of combinations that are possible from applying each recommended element family is extremely large. It will also increase by many degrees of magnitude if we consider the number and characteristics of particular actuators. Different techniques are available in order to reduce this ample selection, reaching a meaningful number of relevant choices.

A popular method to find a given solution is the application of multi-objective optimization, using different algorithm variants (Yezioro and Capeluto 2007; Stadler et al. 2016). However, the technique still remains complex, with restricted application by experts in the subject. Although its intention is to provide very narrow result sets, an optimization procedure based exclusively on performance of one or a few aspects, can severely restrict design variety, choices or project implementation, or produce a conflict with stated objectives (Narzisi et al. 2006). It can also prove difficult to use when assigning values to weighted factors that have a vague definition, but help decide on one alternative over the other.

Examples include environmental benefits derived by the use of a given technology through its reduction on fuel energy generation, subjective appreciations concerning comfort given by an envelope combination, contractor expertise in using innovative constructive methods or the added-value of a new building. Some of these "hard to quantify factors" end up being criteria used by design teams in the real world to decide on one alternative over the other.

Other quantitative methods to select alternatives from a large set of possible solutions include ranking and elimination with preference (Roy 1991), rough set theory (Pawlak 1982), as well as database lookup (which is different to evolutionary algorithms, by using previously calculated sets of combinations in order to select suitable solutions). Readers are invited to research further on any methods that pose any interest for their particular purpose, with emphasis on those that provide a varied set of solutions.

Returning to our example for a new office building, let us suppose that, due to the large number of options available the design team wishes to narrow its choices in an informed way. Due to the short time available to reach a preliminary design, requires to have definite options that are candidates for high marks in the local energy certification system.

Based on the principles discussed in Chap. 4, a suitable solution is the use of design-suggestion tools. Although they are not numerous, they can provide a series of alternatives for evaluation by designers that comply with given characteristics. In this case, intelligent envelope elements that comply with an energy code or certification. An example of such tool is NewFacades (Ochoa and Capeluto 2009). It generates a series of alternatives based on initial conditions and desired design intentions, and returns a series of different proposals for intelligent façade combinations that comply with a chosen energy code or certification.

For the main orientation, the six of the best performing combinations in terms of energy consumption are summarized in Table 5.2.

In the table, three combinations are shown for West, as originally proposed in the brief. The output also shows three options on the South, chosen in order to explore what-if scenarios, for example what could happen if the design was entirely guided by climatic principles. In this way, the designer is exposed to additional 'out of the box' options and a degree of variety is introduced.

The software model NewFacades provides indications for insulation levels of the opaque component, adequate window sizing according to the studied location, light controls, and suggestions for fixed light redirection elements to be incorporated on the building envelope. It also provides recommended glazing properties, but for brevity the table only mentions a basic denomination for each type. Also very important, NewFacades provides as a result, input files that can be used in a whole-building energy simulation program. Thus, the design team does not need to compose any of those files, which can be a time-consuming process.

The next section presents additional considerations on how decisions would be reached, based on the available design options.

5.2 Considerations for Alternative Selection in a New Project

This section shows a sample pathway towards deciding on a suitable combination of envelope elements, while satisfying initial design considerations, energy performance and visual comfort. We are assuming that to reach these decisions, whole-building energy calculations and comparisons have been made for all the combinations shown in Table 5.2. Results show that alternatives, within the same orientation, change according to design variables, but changes between West and South cases are more significant due to the fundamental design decision of window orientation.

As indicated in the sample brief of Sect. 5.1 for choosing certain elements, the desired project image should convey prestige and novelty through its intelligent envelope. This makes external elements suitable, as their reactions and changes can

Table 5.2 Sample listing of possible alternatives for a new building, generated by a design-suggestion program (Ochoa and Capeluto 2009)

CASE A—West	CASE B—South
Alternative 1	
U: 0.6	U: 0.6
WS: 25 %	WS: 25 %
G: Green tint double glazing No lightshelf	G: Green tint double glazing External lightshelf
S: External vertical louver w/sensor	S: External horizontal blind 2″ wide wood w/sensor
V: Night ventilation-windows+fan w/sensor	V: Night ventilation-windows+fan w/sensor
LC: Dimmer	LC: Dimmer
Alternative 2	
U: 0.6	U: 0.6
WS: 25 %	WS: 25 %
G: Low-E External lightshelf	G: Low-E No lightshelf
S: Internal rolling shade medium reflectance, high insulation w/sensor	S: Internal rolling shade medium reflectance, high insulation w/sensor
V: Night ventilation-windows+fan w/timer	V: Night ventilation-windows+fan w/timer
LC: All lights turn on/off when required level is reached	LC: All lights turn on/off when required level is reached
Alternative 3	
U: 0.6	U: 0.6
WS: 25 %	WS: 25 %
G: Clear double glazing Internal lightshelf	G: Clear double glazing No lightshelf
S: Internal rolling shade high reflectance, low solar transmittance w/sensor	S: Internal horizontal blind 1″ wide plastic low reflectance w/sensor
V: Night ventilation-windows+fan w/timer	V: Night ventilation-windows+fan w/timer
LC: All lights turn on/off when required level is reached	LC: All lights turn on/off when required level is reached

Note Alternatives contain solar water heater as mandated by local building code
Codes U U-value opaque wall (W/m^2-C), *WS* window size as percentage of total floor area, *G* glazing, *S* shading, *V* ventilation, *LC* light control

be seen from the outside and identified as an outstanding feature of the building. However is still important to investigate also different design directions and their implications.

On the economic side, however, the payback period has to be acceptable for investors. Using this guideline, solutions that have well-known performance and maintenance characteristics should be sought, such as automated blinds and outer louvers. At the time of writing, these systems generate lower maintenance and replacement costs than similar options such as rolling shades, assuming that they are properly installed and commissioned.

External elements may also contribute to achieve the original intention of West-facing views to the main orientation, although they would have slight interruptions caused by the blinds. Vertical opaque shade elements also manage better solar radiation for the West especially if their angle of obstruction can be adjusted dynamically.

In terms of cooling, a simplified method in this climate that can bring significant savings, which becomes the dominant requirement, is allowing natural ventilation and the selection of automated night ventilation to cool down the thermal mass of the building.

The choice of glazing may be affected by the complete shading solution, allowing daylight use while avoiding overheating and glare. On the other hand, the building facade might offer, from the outside, different types of transparency degrees according to the individualized shading levels applied by users inside. These changes might be acceptable or not according to aesthetic decisions from the project owner or the design team. This is also a consideration to be taken into account for the external elements that might need to be modulated according to the usage partitions.

The South orientation (Case B) is easier to control in terms of solar and glare protection through external or internal shading elements, even though it might not offer the desired views that are sought by the project developer. Choices for lighting systems in all alternatives will depend on economic viability of sensors and type of lamps being installed.

Based on the results, the designer might consider refining the initial scheme according to what is allowed by project conditions: Simpler changes would include exploring further external/internal devices or increase device sophistication. Making a final verdict on a single alternative will depend on economic feasibility, return of investment and life cycle analysis besides aesthetic considerations.

5.3 Scenario Two: Residential Building Retrofit

Specific prospects and challenges are opened in the field of energy retrofit of residential buildings. Research is needed to develop new systems that correctly apply energy strategies and achieve high savings, making existing buildings closer to achieve a near zero-energy ideal (Li et al. 2013). In this way, new options will be available for national construction policies that deal with energy reduction on existing building stock (Uihlein and Eder 2010). Energy retrofit for residential buildings not only reduces energy use and greenhouse gas emissions. It also provides a chance to occupants in social housing to improve their thermal comfort conditions and deal with energy poverty. More than usual, older structures lack updated conditioning equipment, resulting in a lower perceived economic value when compared to new energy-efficient buildings (Popescu et al. 2012).

Since the number of affected buildings is extremely large, it is necessary to take a different approach from traditional renovation. In this way, energy savings and emission reduction results will be obtained in a significant time frame. Renovation methods that use a traditional method, such as envelope demolition and rebuilding, have extended execution times that result in prolonged disruption of the lifestyle of those inhabiting the project.

On the other hand, using prefabricated elements is faster and reduces construction times. The implementation of prefabricated elements to achieve energy savings for existing residential buildings has been suggested as an international policy. It is a viable solution that shortens the process (Uihlein and Eder 2008) and reduces costs. It also allows focusing on punctual actions, which are more appropriate. A significant option is to start from the facade, since residential envelopes account for 20–30 % of total energy consumption (Dall'O et al. 2012).

To describe the building retrofit scenario, a theoretical project will be demonstrated for a residential apartment building located in Athens, Greece. In contrast to the previous case study, more limitations (that are likely to be found by designers who deal with existing buildings) will be considered related to this project. They can include stylistic, conservational, morphological financial and constructive ones. Additionally, a prefabricated envelope system will be used. It must also be noted that retrofitting building envelopes with prefabricated systems for energy performance, is still a field where new developments are taking place.

This example refers to a 5-storey residential building in the urban zone of Athens, Greece (geographical coordinates 37° 57′ N 23° 42′ E). The weather is also hot-humid Mediterranean with mild winters. Precipitation averages 414 mm annually, with some snowfalls recorded. Temperatures can average 9 °C during January, and surpass 40 °C during summer months. An additional phenomenon, present in the city, is that of the urban heat island due to its topographic features (Founda 2011). Electronic climatic data for the location is also available as TMY format.

The building under study was built in the 1970s, before the enactment of the first energy construction codes. It has three free facades (North, West and South) that will be renovated, but in the example only the South façade will be discussed. The overall condition is that of a building with low maintenance showing signs of deterioration. This will influence the amount of resources left available for the energy upgrading itself. Thus, it is important to weight energy strategies and envelope elements that can reach comfort levels, but without burdening other aspects of the building renovation.

In the case being considered, the design team wants to implement some degree of participative design. For this purpose, coordination meetings have been held with the apartment owners. The design team has found from these consultations that owners want their building to reflect novelty, but urban guidelines must be respected concerning some traditional elements such as terraces and matte surfaces.

Since cost is an important component in social housing projects, a socioeconomic study is supposed to have been carried out. In this case the most suitable economic model is that of a low initial investment, in order for residents to pay smaller fees.

(a) **Collection of relevant information for whole-building energy evaluation**

Although initial information has been mentioned in Sect. 5.1, there is more data that requires being collected previous to energy analysis for retrofit. With an existing building, however, there are always doubts on the type and amount of information that has to be collected for adequate project characterization in whole-building modelling programs.

Intuitively, we can guess that collecting down to the smallest detail is perhaps unrealistic and unhelpful. Some information might never be known, such as inhabitants' patterns of use for certain electrical appliances. Other data might not be available, based on privacy grounds or refusal of inhabitants to cooperate. On the other hand, missing out on important features can result in large deviations between expected and real energy savings. Consequently, this dissimilarity can affect the accuracy of financial predictions that are based on how much energy is *not* being used.

Providing an importance rating for parameters that need to be used for computer simulation has been studied by different authors (Ochoa and Capeluto 2015). Tables 5.3, 5.4 and 5.5 compile initial project variables and their importance to calculate energy consumption for residential buildings through simulation. Table 5.3 makes note of information that has to be collected from site, as well as information that can be acquired from other reliable sources. Examples of the latter include weather data for the climatic region, which can be obtained from the relevant meteorological service or in formats ready for use in dynamic energy simulation programs.

Despite site visits, some information might not be gathered. Table 5.4 points out which data can be assumed, from studying similar buildings or consulting relevant databases. Table 5.5 concludes by pointing out features that the literature is not clear about their influence on energy consumption, and which might require further research before deciding on their importance.

Sample values are given in each one of the tables, which represent most common ranges that can be found in many apartment buildings. Users will need to complete their own tables with specific project information.

(b) **Brief description of the retrofit system**

Some brief details will be given about the prefabricated envelope retrofit system to be applied on this case study. The system was developed in the framework of a European research project and has been previously described (Capeluto and Ochoa 2014).

It has as objective to become an alternative for a prefabricated façade system for energy saving in European apartment buildings that need renovation, addressing European Union policies for energy and emissions reduction. The system is intended to be modular and adaptive, as shown in Fig. 5.2, in order to support different technological and structural options in a single basic unit that will combined to be used in multiple climates. Technologies will be placed inside the basic unit according to the needs imposed by the local climate and project conditions.

Table 5.3 Parameters and their importance for energy simulation of residential buildings: Data to be collected on site or from reliable sources

Data type	Parameter	Possible ranges	Importance
Location	Geographical location	Latitude, longitude, altitude, HDD, CDD	Very high
Geometry and size	Orientation	16 directions	High
	Proportion plan	1:1, 2:1, 1:2	Medium to very high
	Number external walls	1 to 3 (4 possible but less common)	Medium to high
	External wall area	Dependent on floor-to-floor height	Medium to high
	Apartment location	Ground floor to top	High
	Building plan shape	Slab, cross, L, H, etc.	For self-shading
Existing façade	External wall thickness	0.10–0.30 m	Low
	Wall composition (including insulation)	Variable	High
	Color existing wall (albedo)	Dark to very bright	Very low (with good insulation)
Window	Window area	10–30 % floor area	Very high
	Glazing	Single, double, triple (for glass materials)	Medium to high
	Frame material	Metal, wood, plastic, w/insulation	Low to medium
	External shading	None, movable, fixed (also terraces, balconies)	Medium to high
	Internal shading	None, thin curtains, thick curtains, blinds	Low to medium

Another objective of the prefabricated modular system is to reduce energy retrofit costs through mass production, enabling a reduced market price and shorter non-intrusive installation process.

Among the options being offered, there are modules that can be placed on window openings for shading and glazing upgrading. Basic modules for the opaque section of the facade allow EPS insulation with different color options for the outside layer (which can be changed to PCM if required), and ventilation. Optional modules for opaque areas of the facade include building-integrated photovoltaic panels (BIPVs) for self-electricity generation, and green facades for passive thermal mass cooling.

Therefore, the system provides flexibility to combine both opaque and window sector placements if required. When existing windows need to be removed, opaque wall modules can be anchored without need to demolish existing walls. It also gives the possibility to use a single technology type in all the facade or to combine a number of them according to specifics needs.

Table 5.4 Parameters and their importance for energy simulation of residential buildings: Data that can be assumed from site visit or similar building types

Data type	Parameter	Possible ranges	Importance
Infiltration	Infiltration rate (ach)	0.1–1.5 ach (tests needed on field)	Medium to very high (climate and construction dependent)
Ventilation	Natural ventilation	Always on, seasonal, none	Medium
	Night ventilation	Seasonal, none	Medium
HVAC	Mechanical ventilation	Present, absent	Medium to high
	Mechanical ventilation capacity	1–2 air changes per hour (ach)	Required for indoor air quality
	Mechanical ventilation efficiency	Efficiency fan	Medium to high (climate dep.)
	Cooling system and its efficiency	Different household examples	Very high
	Heating system and its efficiency	Different household examples	Very high
	HVAC control system	Automatic, manual, timer	Medium to high
Setpoints	Cooling temperature setpoint (day-night)	25–27 °C (country dependent)	High
	Heating temperature setpoint (day-night)	21–17 °C (country dependent)	High
Loads	People	0–6 persons	High
	Occupancy schedules	Variable	High
	Equipment	Variable	High
	Equipment schedule	0–100 % usage	High
Internal thermal mass	Thermal mass internal elements	0 to full thermal mass	Low to medium (higher thermal mass envelope)
Slabs	Slab composition	Variable	Low
	Slab thickness	0.15–0.30 m	Low
	Contribution to energy transfer	Adiabatic, part of thermal mass	Low to medium
Internal walls	Composition walls	Variable	Low
	Thickness	0.10–0.30 m	Low
	Contribution to energy transfer	Adiabatic, part of thermal mass	Low to medium

(continued)

Table 5.4 (continued)

Data type	Parameter	Possible ranges	Importance
Surroundings	Neighboring buildings and density	Contiguous, spaced, isolated	Medium
	Immediate vegetation and height/type	Grass, bushes, low trees, high trees	Low to medium
	Surfaces' material	Variable	Low to medium
	Surfaces' reflectivity	Dark to very bright	Low to medium
Electric lighting	Lighting type and amount	Incandescent, fluorescent, LED, "energy saving"	High
	Lighting schedules	Variable	High
	Lighting user profile	Active, average, aging	High
	Lighting control	On/Off, dimmer	Low (housing)

Table 5.5 Parameters and their importance for energy simulation of residential buildings: Data with high variation but no definite knowledge on its influence for building simulation

Data type	Parameter	Possible ranges	Importance
User features	Residents' age	N/a	N/a
	Tenancy	Renting, property	N/a
	Energy payment plan	Meter reading, averaged, included in rent	N/a
	Ratio area conditioned by users	N/a	N/a
	Resident's income	N/a	N/a
Other building features	Presence/absence of basement/attic	Attic, basement, both, none	N/a

The retrofit system also addresses the shortcomings of traditional prefabricated systems by offering a selection of different material finishes. These finishes are placed as the external layer of insulation units.

(c) **Climatic strategies allowed by the system**

Although theoretically new technological units can be developed, in this example the number of climate strategies that can be applied with the prefabricated retrofit system are already defined, due to number of available options for each module. However, correct strategies and preferred combinations need to be defined for each location, in order to select modules relevant to the particular project requirements.

During the course of research to develop the system, climate strategies provided by the actuators were determined for three broad climate areas in Europe. This was

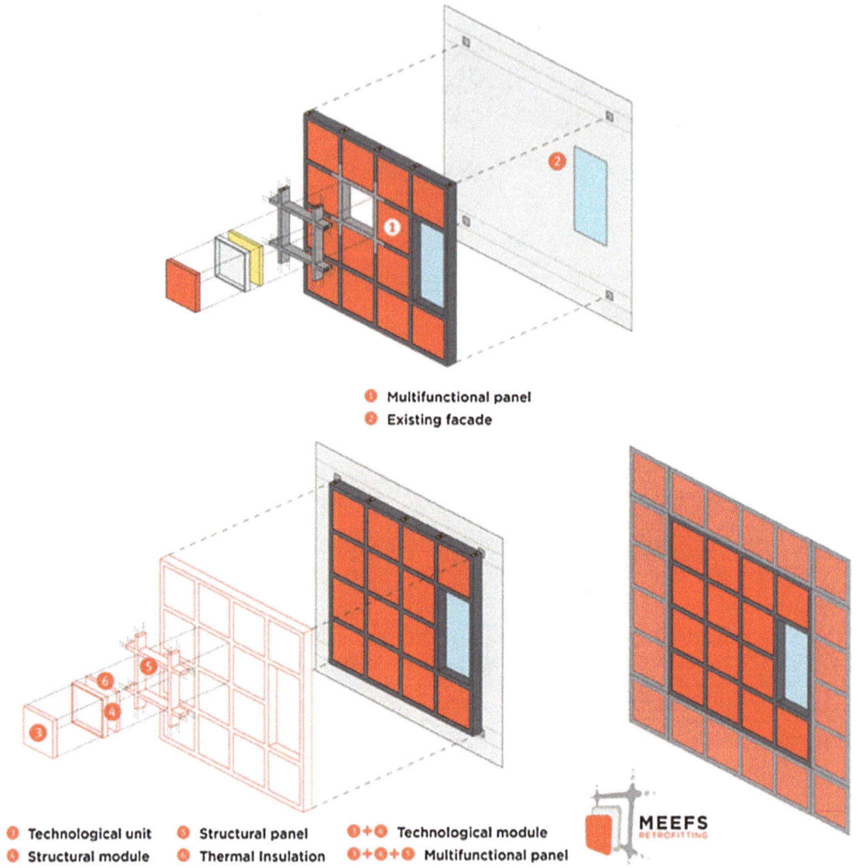

Fig. 5.2 Schematic view of the multifunctional energy efficient façade system ("MEEFS") for façade retrofitting. *Image credit* MEEFS project

achieved through general energy simulations for each climate type, using a base residential module with typical constructive characteristics for different cities within each climate zone.

The results of applying single strategies and combinations were summarized synoptically for all four main orientations and thirteen representative studied cities. An example for the South orientation is shown in Fig. 5.3. It shows a hierarchic organization of strategies according to the amount of predicted energy savings they can achieve in a given city. Up to four rankings of each strategy are given. It also presents the impact of using a single strategy or a combination of two of them.

The figure is intended to be used as a time-saving tool to facilitate decision-making when considering the selection of a set of modules and their technologies. With this tool, designers can choose according to the best performing strategies while considering additional factors such as priorities in terms of high energy

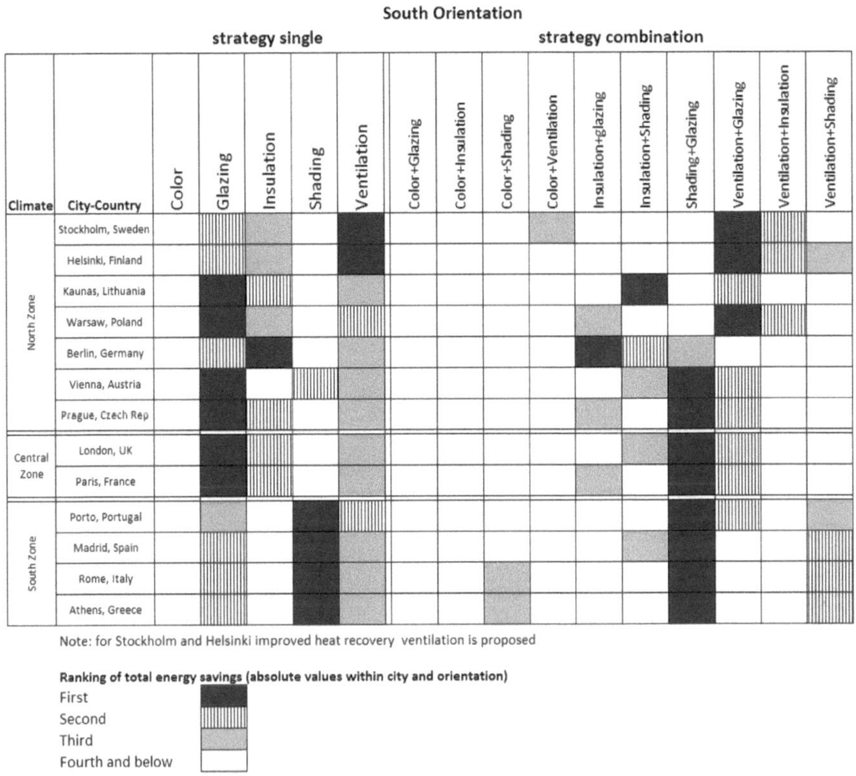

Fig. 5.3 Synoptic table for strategy ranking for different cities and climate types using a prefabricated energy retrofit system, South orientation

saving performance, budget and payback, ending with stated architectural intentions for the project.

For Athens, the strategies that can be applied using the described retrofit system can be found in Fig. 5.2 under the section "South Zone". They indicate that for its climate type, shading is the single strategy provides the highest energy savings, followed by ventilation and glazing replacement. When two strategies are combined, then shading and glazing is the most effective strategy, followed by ventilation and glazing change.

(d) **Selection process**

Returning to our case study, we provide an example on the use of available project information to infer numerical values that are usable in whole-building energy simulation, and in turn, to start analysis and take decisions. Based on the original construction year of the project (1970s), and the degree of maintenance of the façade (showing signs of deterioration), a series of characteristics can be determined by consulting relevant building databases. Many of them were compiled for different European research projects.

Examining them indicates that, buildings in Athens with very similar characteristics to the case study, have an average U-value of 2.2 W/m^2-C for the opaque component, and infiltration value of 1.0 ach, among other data (Susref 2016). On the other hand, the recommended minimum façade U-value that must be complied with is 0.6 W/m^2-C, according to the 2010 version of the Energy Performance of Buildings Directive (EPBD) for Greece.

These values were used, in addition to other data from the same reference characterizing a typical façade that matches the properties of the case study. Through whole-building energy simulations made for the conditions described above, yearly energy demand for heating and cooling was found to be around 65 kWh/m^2 for South-facing apartments without any technology being applied.

The energy consumption result is consistent with the findings for the basecase and the technological combinations, which are shown in Fig. 5.4. The figure shows the influence on heating and cooling consumption of applying one and two strategies of the prefabricated façade retrofit system for the South orientation in the basecase for Athens. In this way, we can use these results for checking different design alternatives and finding the most suitable combinations based on project requisites.

In order to allow stakeholders taking construction decisions, such as building and urban designers, to consider other design directions fulfilling project requirements, it is convenient to introduce variety in the final filtering. In the case of architectural goals, an option fulfilling a different direction but with similar energy and economic performance may be considered. It can be added to the first three proposed options that do satisfy the requested architectural intentions, in order to be considered by the designer. For the specific case study, one of these solutions will correspond to the category of a "modern" design intention.

A sample procedure to select combinations based on energy performance and economic factors is presented in Table 5.6. It is organized by the number of modules used, and priorities in terms of high energy performance or traditional design with low initial cost. Financial information in the table was calculated by analysts involved in the life-cycle and cost aspects of the research project (Kürümlüoglu et al. 2009).

The number of modules in the table corresponds to the number of technology options being placed for the particular combination. They are applied to all the relevant locations in the façade, without causing any functional interference between each other. Energy savings are expressed as a percentage relative to the current situation (before retrofit).

The financial cost ratio of solutions (CS/CI) is calculated as the retrofit cost using the modules (CS) over the cost of traditional energy retrofit using insulation upgrading only (CI). Expressing this figure as a ratio enables readers to compare different solutions with relative independence of market and time. Therefore, a higher CS/CI ratio means a more expensive solution when compared to traditional energy retrofit.

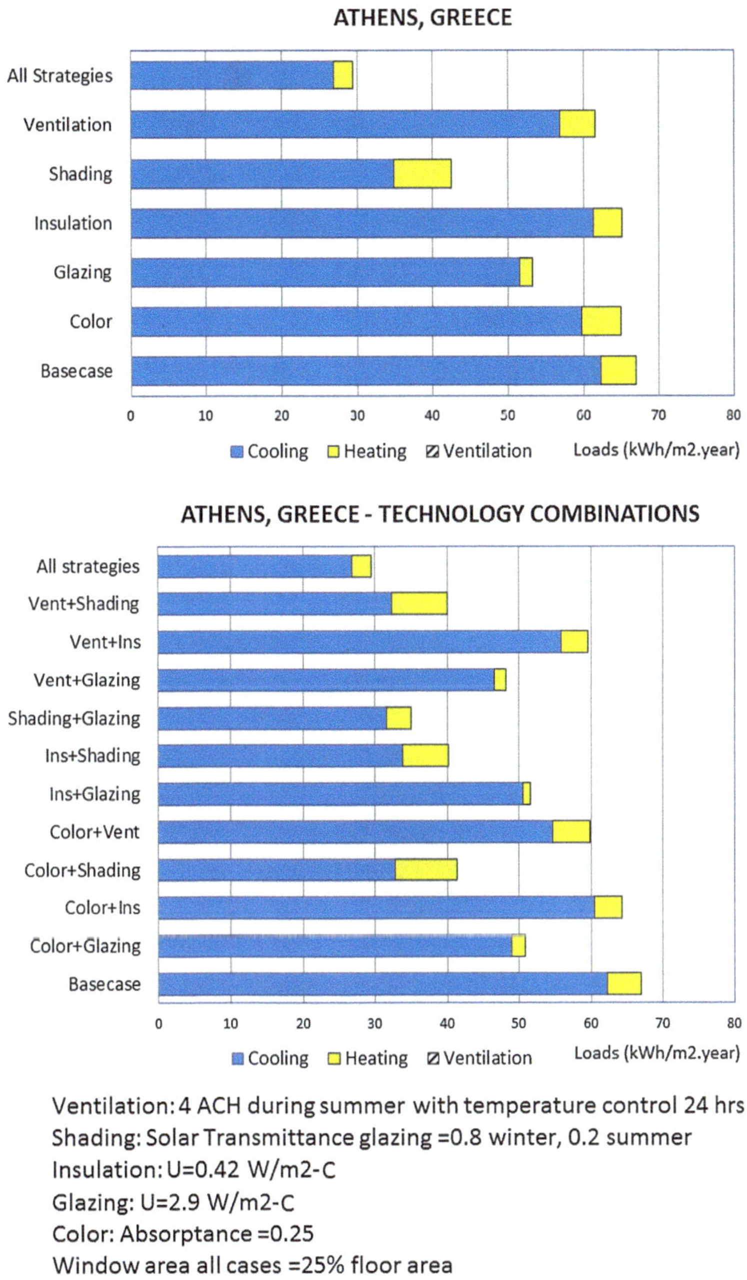

Fig. 5.4 Energy consumption for basecase and strategy combinations applying a modular façade energy retrofit system applied in Athens, Greece, South orientation

Table 5.6 Summary of selected solutions suggested by the methodology for a hypothetical case of a South façade in Athens, Greece

Modules used	Priority	Option	Combination	Design intention	CS/CI	IRR	Energy saving %
2	High energy performance only	A	Ins+DG	Traditional	1.05	1.30	41.19
		B	BIPV+DG[a]	Modern	1.17	1.23	39.78
		C	Ventilated+ DG ins	Modern	1.54	0.49	40.14
		D	Green+DG	Traditional	1.60	0.47	22.01
	Low initial cost and traditional design	E	Ins+DG	Traditional	1.05	1.30	41.19
		F	BIPV+DG[a]	Modern	1.17	1.23	39.78
		G	Green+DG	Traditional	1.60	0.47	22.01
		H	SO+DG sol	Traditional	1.64	−0.60	17.57
3	High energy performance only	I	Green+SO+ DG ins	Traditional	3.33	1.06	44.27
		J	Ins+SO+DG ins	Traditional	1.90	0.50	42.67
		K	Vent+SO+ DG ins	Modern	2.39	0.37	41.15
		L	Ins+SS+DG ins	Traditional	1.90	0.33	36.23
	Low initial cost and traditional design	M	Ins+SO+DG ins	Traditional	1.90	0.50	42.67
		N	Ins+SS+DG ins	Traditional	1.90	0.33	36.23
		O	Vent+SO+ DG	Modern	2.39	0.78	41.12
		P	Green+SO+ DG ins	Traditional	3.33	1.06	44.27
4	High energy performance only	Q	Ins+Green+ SO+DG ins	Traditional	3.62	0.88	43.31
		R	Vent+Green+ SO+DG	Modern	4.12	0.88	41.57
		S	Ins+Green+ SS+DG ins	Traditional	3.62	0.53	36.96
		T	Ins+Green+ SS+DG sol	Modern	3.62	0.36	33.98
	Low initial cost and traditional design	U	Ins+Green+ SO+DG ins	Traditional	3.62	0.88	43.31
		V	Ins+Green+ SO+DG	Traditional	3.62	0.88	43.21

(continued)

Table 5.6 (continued)

Modules used	Priority	Option	Combination	Design intention	CS/CI	IRR	Energy saving %
		W	Ins+Green+ SS+DG ins	Traditional	3.62	0.53	36.96
		X	Vent+Green+ SO+DG ins	Modern	4.12	0.88	41.60

CS/CI Initial costs relative to placing insulation only over opaque section of the façade. *IRR* Investment recovery rate. Saving % relative to basecase
Ins EPS Insulation
Green Modular green facade with irrigation system
SO Aluminium solar overhang, 0.50 cm
DG clear double pane glazing
DG ins thermally insulating double pane glazing
Vent Passive ventilation module with operable slats
BIPV Facade photovoltaic panels[a]
SS Aluminium operable solar slats
DG sol solar selective double pane glazing
[a]Photovoltaic electric output redirected to the grid, calculated at 127.39 kWh/m^2-year

The investment recovery rate (IRR) is defined as the retrofit cost of the combination during its first year divided over the retrofit cost (or gain) until twenty years of operation. The amortization period of 20 years is used as a common reference time frame for analysis as a suitable mortgage period. It is a complex figure that considers: total energy savings, power generation (if any), maintenance and replacement, as well as social factors. Any quantifiable indirect benefits to society brought by these technologies were considered as social factors, such as: improved comfort conditions, benefits to health, appreciation of property, etc. This can enable choosing a given technology with high initial costs but also providing high-performance yield. This means that a higher IRR is a solution where costs can be recovered faster due to the energy savings and benefits brought by the solution.

From the table it can be seen that, in some cases, the choice for low initial cost also coincides with high energy savings and high return rate. Since there were no specific budget limits on the definition of the case study, users would be presented with additional information for the array of solutions ranging from 1, 2, 3 and 4 modules. In general, solutions with three or four technological modules have higher initial cost than single or double technological modules on the façade.

The analysis of neighboring obstructions, such as height, geometry and materials, is needed to complement technology selection. It is also required to determine the placement of these technologies within the façade. An example includes solar based technologies such as BIPVs, where mutual building shading can reduce estimated performance and profitability. The detail presented by this analysis provides uniqueness suitable to each case, while still being applicable to a large number of situations.

Further explorations on the options would improve and optimize details for specification and energy savings. Results might need to be refined in more advanced design stages. As final design and usage data become clearer, it will aid in verifying projected and actual energy savings.

5.4 Considerations for Element Selection in a Retrofit Project

Initial calculations assume that simultaneous use of all strategies is possible without interference from each other, and thus produce the highest energy savings. However, many projects will require solutions with other configurations due to different constraints. Some of them include: limited existing façade area unable to accommodate all technologies, partial shade from external obstructions, or unforeseeable occupant attitudes towards new heating or cooling equipment.

Other limitations for the use of certain technologies include available budget, aesthetical constrains, or local regulations. One example can be found in some historical centers, that due to their materials and aesthetic value, local authorities forbid the use of large glazed or shiny surfaces. The intention of this limitation is to maintain a unified aspect, becoming an additional challenge for the retrofit project. In this case, and for the prefabricated system, Fig. 5.2 can guide designers on the most suitable strategies or their combinations.

Another practical assumption that has been made is the existence of a communal agreement on façade use to place modules and their technologies considering the building as a whole. Although the aspect is unrelated to energy performance or economic analysis, it is important to clarify it. According to different ownership and maintenance models present in multi-level buildings, reaching such agreement will allow the project to continue beyond any potential legal disputes arising from that issue.

It was assumed for the preliminary analysis, that the layout of the selected technologies was optimal within the façade. Specific site analysis such as solar irradiation on the facade, wind directions, existing neighboring buildings and topography are needed in order to know their exact positions and influence. The study can be done using different design tools mentioned in Chap. 4, in order to reach a conceptual design based on the information that they can provide. The complexity of selecting suitable energy strategies and technology increases when considering additional economic, legal, structural and aesthetic factors at the same time.

Economic considerations form part of the decisive factors to carry out a retrofit project, particularly in social housing. Economic feasibility, in order to be operative, needs to be updated continuously for initial cost and payback calculations of technologies, reflecting the dynamics of each market. If novel technological developments can provide lower initial costs and faster payback periods, then new recovery rates should be considered. Different variables must be taken into account, to generalize for all markets the economic impact of environmental benefits brought by energy saving technologies.

Technologies with higher accumulated financial returns, greater comfort benefits and lower initial costs will be favored to ensure project completion and commissioning. Therefore, financial appraisal cannot be limited exclusively to initial expenditure to achieve project completion.

References

Bitan A, Rubin S (2000) Climatic atlas of Israel for physical and environmental planning and design, 3rd edn. Tel Aviv University Geographical Society of Israel and Ministry of Energy and Infrastructure, Tel Aviv

Capeluto IG (2003) The influence of the urban environment on the availability of daylighting in office buildings in Israel. Build Environ 38:745–752

Capeluto IG, Ochoa CE (2014) Simulation-based method to determine climatic energy strategies of an adaptable building retrofit façade system. Energy 76:375–384

Dall'O G, Galante A, Pasetti G (2012) A methodology for evaluating the potential energy savings of retrofitting residential building stocks. Sustain Cities Soc 4:12–21

Founda D (2011) Evolution of the air temperature in Athens and evidence of climatic change: a review. Adv Build Energy Res 5:7–41

Kürümlüoglu M, Schubert M, Eckstein H (2009) Risikomanagement bei der Einführung eines PLM-Systems. ZWF Zeitschrift fur Wirtschaftlichen Fabrikbetrieb November 2009:982–987 (In German)

Li DHW, Yang L, Lam JC (2013) Zero energy buildings and sustainable development implications-a review. Energy 54:1–10

Meteonorm (2016) Meteonorm software. http://www.meteonorm.com/. Accessed May 2016

Milne M (2016) Climate Consultant v6.0. Available at http://www.energy-design-tools.aud.ucla.edu/. Accessed Mar 2016

Narzisi G, Mysore V, Mishra B (2006) Multi-objective evolutionary optimization of agent-based models: An application to emergency response planning. In: Kovalerchuk B (ed) Proceedings of the 2nd IASTED International Conference on Computational Intelligence, CI 2006. November 20–22. San Francisco, California, USA

Ochoa CE, Capeluto IG (2006) Evaluating visual comfort and performance of three natural lighting systems for deep office buildings in highly luminous climates. Build Environ 41:1128–1135

Ochoa CE, Capeluto IG (2009) Advice tool for early design stages of intelligent facades based on energy and visual comfort approach. Energy Build 41:480–488

Ochoa CE, Capeluto IG (2015) Decision methodology for the development of an expert system applied in an adaptable energy retrofit façade system for residential buildings. Renew Energy 78:498–508

Pawlak Z (1982) Rough sets. Int J Inf Comput Sci 11:341–356

Popescu D, Bienert S, Schützenhofer C, Boazu R (2012) Impact of energy efficiency measures on the economic value of buildings. Appl Energy 89:454–463

Roy B (1991) The outranking approach and the foundations of ELECTRE methods. Theor Decisions 31:49–73

Stadler P, Ashouri A, Marechal F (2016) Model-based optimization of distributed and renewable energy systems in buildings. Energy Build 120:103–113

Susref project (2016) Susref website http://cic.vtt.fi/susref/. Accessed May 2016

Uihlein A, Eder P (2008) Towards additional policies to improve the environmental performance of buildings. Report number JRC-IPTS, EUR 23775 EN. European Commission, Office for Official Publications of the European Communities, Luxembourg

Uihlein A, Eder P (2010) Policy options towards an energy efficient residential building stock in the EU-27. Energy Build 42:791–798

Yezioro A, Capeluto IG (2007) A model for the energetic-economic optimization of office buildings. Archit Sci Rev 50:331–339

Conclusions—The Intelligent Envelope, Where To?

Throughout this book, we have shown that true intelligent building envelopes can only be the result of an informed design process that itself requires intelligence. Yet there are some requirements and differences for this process to be successful, distinguishing itself from the "traditional" architectural design activity.

The intelligent design process for high-performance building envelopes is a conscious and well-directed effort, which must start very early in the design process. During all its stages it must consider smart architectural design decisions at the building level, as well as passive design strategies. The process understands and applies the procedures behind obtaining a low energy design, while at the same time achieving maximal occupant comfort. Concrete and organized selection approaches direct the designer towards selecting the most suitable elements, with the aim of executing in a coordinated fashion distinct climatic strategies on the outer layers of a building, those that mediate between inside and outside. These strategies are adaptable to the challenges posed by the environment, following a *method* similar to how living beings adapt within their environment and that can be considered intelligent. This process follows the basic steps of perception, reasoning and action.

As such, we can talk of *artificial intelligence* being placed on inanimate objects intended to act systematically when confronted with a variety of environmental challenges and conditions. Such effort strives for logical consequences triggered by the surrounding context and which bring an improvement compared to the initial condition. For high-performance intelligent building envelopes, these actions should bring objective and appreciative benefits to the occupants. In this context, it must be clarified that not all elements chosen to undertake climate strategies need to have artificial cognitive capabilities, as it might add unwanted complexity to the strategy they are solving and sometimes correct early design decision avoid complex late solutions. Nevertheless, sometimes there are advantages and added benefits if selected elements are provided of such capabilities mainly if they are based on conscious design decisions.

A true intelligent envelope requires components that have been *chosen intelligently*. It must be considered that facades and roofs cannot be provided with just a collection of "smart" elements placed together, and then somehow expected to perform adequately. Even those components presenting smart materials and with

G. Capeluto and C.E. Ochoa, *Intelligent Envelopes for High-Performance Buildings*, Green Energy and Technology, DOI 10.1007/978-3-319-39255-4

the highest degree of mechanical or electronical complexity require to be pre-programmed with adequate responses to the different challenges they are going to meet. This makes climatic-oriented intelligent design essential, in order to take advantage of the wide variety of components available for intelligent envelopes: passive systems, smart materials, robotic actuators, sensors, etc.

It is their well-studied combination, rather than the prevalence of one family of components over the other, which enriches design. The results of careful consideration of different strategies and the means to achieve them, results in a balanced responsive building envelope. Through responses that reflect the intelligent design process, it is able to cleverly sense, reason and respond to a series of requirements that are ultimately posed by users internally and by the environment externally.

It must be remembered that for an adequate design activity and to organize it more efficiently, the intelligence process can be understood from a functional point of view. It has been mentioned that three main linear stages are usually considered: perception, reasoning and action. They are mirrored, in the realm of inanimate objects, by sensors, processors and actuators. The relationship between intelligent process functions, and main element families carrying them out, is one of the fundamental blocks for envelope design analysis using an intelligence focus.

The other, not less important fundamental block, is the use of responsive climatic design, making intelligent envelope design particular and unique for every location. It requires study of weather characteristics related to human comfort, in order to formulate adequate response strategies and search for the most suitable design solutions and actuators that can meet them according to project conditions. The crafting and implementation of these responses provides the potential of making an intelligent building as smart as its designers.

The current state of the art in terms of resources available for intelligent envelope design provides a good starting point. A variety of built and occupied examples are available for study, constructed during different decades and in a variety of locations. Successes, pitfalls and constructive implementation can be learned; correlating how design philosophies have operated on the final product, its performance, and how users have interacted with the different degrees of automation and cognition offered by these buildings.

Similarly, those undertaking the design of an intelligent building envelope can benefit from modern climatic design tools. They have taken advantage of computer hardware developments, offering speed in the analysis of a variety of complex variables and their interactions. Design tools can be found as simple manual tools, stand-alone software programs or as complements to well-known drafting and building information modelling suites. This has expanded the means of studying a project and performing corrective operations to the agreed design well before construction. Nonetheless, what these tools still do not always offer is a user-friendly approach to their use according to each design stage.

At the time of writing, although more powerful, tools that can help design intelligent envelopes remain complex as when they were first conceived some decades ago. They generally evaluate finished alternatives rather than suggest design directions, and their correct usage can mainly be done only through specialists. Our aim for this is not to criticize tool developers, but rather to encourage development of tools that will allow asking relevant questions ("what if" scenarios), and in return provide meaningful information according to the needs of each design stage, increasing the performance of future intelligent envelopes. It is in the early design stages when architects are tinkering with ideas, and exactly then is the moment when performance of a future building is being defined.

Suggesting a series of solutions or possible promising design directions is also important for real-life architectural design. Few times designers encounter a linear problem that has a unique solution. A series of intricate factors must be weighted to design an already complex and important building segment as the building envelope. Analysis has to be made of the project site and its obstructions, examination must be done on the paths to choose the most appropriate energy strategies, and how to select possible sets of solutions for the envelope components according to design priorities. These priorities in many occasions are not decided by the design team but imposed upon them—cost, urban guidelines, technological availability on site, and owner/user acceptance are among some of them.

It can also be hoped that analytical design tools will also evolve from accuracy to applicability, in particular during the early stages of the architectural design process. Real-time evaluation of diverse parameters during actual design is a goal to be achieved, although some current BIM interfaces aim to do part of this evaluation with defined designs at different stages.

Design education with a focus on the correct implementation of climatic principles and strategies according to each situation is essential. Teaching new generations of architects and engineers these principles will ensure that in the future, proper design of intelligent building and envelopes will be made. It also will trigger further development of advanced tools that can enable the exploration of different novel solutions through an intelligence approach.

In this book we presented two typical cases that can be encountered by designers. The first one engaged the envelope design of a new office building, while the second dealt with energy retrofit of a social housing building envelope. While the former case provides a classical introduction to the factors influencing the design of an intelligent envelope, the latter arises as probable to be encountered more often in the near future by designers. This will be due to the future extension of energy directives to existing buildings, as an effort to cut further on emissions and increase energy efficiency.

Envelope retrofit for energy upgrading using intelligent envelopes is a field that still requires further study. Some obstacles remain for wider application, among them reducing uncertainty factors that are still found between expected savings and actual performance of the refurbished project.

It can be expected that intelligent building envelopes in general will have a much wider implementation in the near future. Due to the nature of the design and construction industry, which goes at a slower pace than other technological developments, the incorporation of automation for energy efficiency in our built environments has taken some time to become acceptable. Many reasons are given against its fuller implementation. Nevertheless, electronic and mechanical systems have more reliability than before, they are more affordable, the number of existing actuators has increased, and there is better understanding of passive system. Proper selection of climatic strategies ensures there is no dependence on one system only.

Any forthcoming directions for the development of high-performance intelligent envelopes will include an improved cognition of perception elements. Processing centers need to handle more detailed information beyond what is available today. They also need to go beyond physical sensing and into direct interaction with human users, providing personalized comfort and better reaction to large numbers of occupants. Certainly, further research has to be done so that the proposed interaction becomes meaningful for an enhanced climatic comfort experience.

To reach this goal, intelligent envelope processors can be expected to become equipped with mathematical prediction models, instead of limiting themselves to the administration of fixed policies. Such improved prediction models would act on short term forecast of conditions and prepare the building for more accurate responses. Actuators, on their part, can benefit from electronic miniaturization and core computing to provide better signal communication with the remaining sections of the building, increasing adaptability and response modes as required by advanced artificial intelligence models.

Further studies are also required to provide new operation modes to well-known actuators such as blinds, shading elements, lighting controls, etc. Other directions for research include new additional explorations in the fields of smart materials and biological synergy, accomplishing a series of functions that currently cannot be carried efficiently with traditional actuators.

Ultimately, intelligent building envelopes for high-performance buildings need to be taken as a viable option among architects and engineers. They can provide answers to a wide variety of solutions, according to specific requirements, and not only for high-end projects. Committing to such objective requires both financial and technological affordability and accessibility. Evolution in the design and construction industries will have a faster pace through adoption of intelligence as a technological improvement. This action will ultimately result in benefits to the wellbeing of end-users, stimulation of economic development, and preservation of our environment in general.

With all that said, we shall remember that intelligent successful buildings cannot be just a group of smart active devices; they must be founded on architectural design that implements basic energy-conscious design principles and combine active and passive intelligence: active features and passive design strategies. The result is to provide maximum occupant comfort by using minimum energy. A truly intelligent building needs to be the product of a design process that incorporates intelligence in all its stages, taking advantage of technological innovations.

Index

Note: Italics indicate specific building projects

G. Capeluto and C.E. Ochoa, *Intelligent Envelopes for High-Performance Buildings*, Green Energy and Technology, DOI 10.1007/978-3-319-39255-4

FSC
www.fsc.org
MIX
Papier | Fördert
gute Waldnutzung
FSC® C083411

Kolibri 360 GmbH
Ferdinand-Jühlke-Straße 7
99095 Erfurt, Deutschland
produktsicherheit@kolibri360.de